Wilbur Olin Atwater, Charles Dayton Woods

Dietary Studies with Reference to the Food of the Negro in

Alabama in 1895 and 1896

Conducted with the Cooperation of the Tuskegee Normal and industrial

Institute and the Agricultural and Mechanical College of Alabama

Wilbur Olin Atwater, Charles Dayton Woods

Dietary Studies with Reference to the Food of the Negro in Alabama in 1895 and 1896
*Conducted with the Cooperation of the Tuskegee Normal and Industrial Institute
and the Agricultural and Mechanical College of Alabama*

ISBN/EAN: 9783337213282

Printed in Europe, USA, Canada, Australia, Japan

Cover: Foto ©berggeist007 / pixelio.de

More available books at **www.hansebooks.com**

299

U. S. DEPARTMENT OF AGRICULTURE.

OFFICE OF EXPERIMENT STATIONS

DIETARY STUDIES

WITH REFERENCE TO THE

FOOD OF THE NEGRO IN ALABAMA

IN

1895 AND 1896.

CONDUCTED WITH THE COOPERATION OF THE TUSKEGEE NORMAL
AND INDUSTRIAL INSTITUTE AND THE AGRICULTURAL
AND MECHANICAL COLLEGE OF ALABAMA.

REPORTED BY

W. O. ATWATER and CHAS. D. WOODS.

WASHINGTON:
GOVERNMENT PRINTING OFFICE.
1897.

LETTER OF TRANSMITTAL.

—

UNITED STATES DEPARTMENT OF AGRICULTURE,
OFFICE OF EXPERIMENT STATIONS,
Washington, D. C., January 15, 1897.

SIR: I have the honor to transmit herewith a report on investigations of the food of the negro in Alabama in 1895 and 1896, prepared by Profs. W. O. Atwater and Chas. D. Woods. These investigations constitute a part of the inquiries made with the funds appropriated by Congress "to enable the Secretary of Agriculture to investigate and report upon the nutritive value of the various articles and commodities used for human food," and were carried on under the supervision of Professor Atwater, special agent in charge of nutrition investigations, in accordance with instructions given by the Director of this Office.

The immediate purpose in conducting an inquiry into the food of the colored population of the Southern States was to obtain information as to the kinds, amounts, and composition of the food materials used. The ulterior purpose was to get light upon the hygienic and pecuniary economy of their diet, its deficiencies, the ways in which it might be improved, and the steps which should be taken to bring about an improvement.

The beginning of such an inquiry was undertaken in the neighborhood of the "Black Belt." The investigations were conducted in cooperation with the Tuskegee Normal and Industrial Institute. This institution was selected because of its relation to the negro, its favorable location, and the especial fitness of its principal, Mr. Booker T. Washington, to aid the Department in such an inquiry. The investigations were instituted by Mr. H. M. Smith, special agent of this Department, and the details of the observations were carried out in large part by Mr. J. W. Hoffman, of the Tuskegee Institute. In the course of the studies visits were made to Tuskegee by Professors Atwater and Woods.

The investigation received essential aid from the Agricultural and Mechanical College of Alabama at Auburn. President W. L. Broun took an especial interest in the work at Tuskegee, and afforded every facility for promoting its successful prosecution. The analyses of the food materials were made under the direction of Prof. B. B. Ross of

3

the college, who is also the chemist of the Alabama Experiment Station. These analyses formed a part of the investigation into the composition of food materials of Alabama, conducted by Professor Ross in cooperation with this Department.

This report is respectfully submitted, with the recommendation that it be published as Bulletin No. 38 of this Office.

Respectfully,   A. C. TRUE,
                                                                  Director.

Hon. J. STERLING MORTON,
    Secretary of Agriculture.

CONTENTS.

ILLUSTRATIONS.

6

DIETARY STUDIES WITH REFERENCE TO THE FOOD OF THE NEGRO IN ALABAMA.

PURPOSE AND PLAN OF THIS BULLETIN.

The purpose of this bulletin is to give an account of studies of the food and nutrition of negroes in the neighborhood of Tuskegee, Ala. The investigation was made during the spring of 1895 and the winter of 1895–96, with the cooperation of the Normal and Agricultural Institute at Tuskegee and of the Alabama Agricultural and Mechanical College at Auburn.

The investigation includes examinations of 20 dietaries of 18 families. Some of the families lived in and close by the village of Tuskegee; the majority were on plantations from 2 to 9 miles distant. Some of the families, especially those in and near the village, showed by their improved conditions of living the noteworthy influence of the Institute and of association with people of intelligence and thrift. The same was true of some of the plantation families. The rest were very much like the ordinary plantation negroes, and were, indeed, selected as typical examples of the great mass of the colored people of this region. A number of the families were on a large plantation where the mortgage system prevails, and the plane of living is a very low one. The people studied represent the gradation from the lower to the higher grade of living which actually exists among the negroes in Alabama, and the observations thus help to illustrate not only the evils under which the colored people live, but some of the phases through which they are passing in their upward progress.

While the main subject is the food and nutrition of typical negro families in a representative district of the South, other observations pertaining to their conditions of life are also reported.

The general plan consisted in visiting each house or cabin from day to day for a period generally of two weeks, weighing the food used by the family, and taking specimens for analysis, notes being made at the same time regarding the people, their dwellings, farms, work, habits, and the like.

COMPOSITION OF ALABAMA FOOD MATERIALS.

In cooperation with this Department, Prof. B. B. Ross, of the Agricultural and Mechanical College of Alabama and the Alabama Experiment Station, made during the spring of 1895 and the winter and spring of 1896 a considerable number of analyses of materials used as food in Alabama. Part of these were the specimens collected in the dietary studies reported beyond. The others were selected elsewhere as typical of the food of the region. The methods used for the analysis of vegetable foods were essentially those of the Association of Official Agricultural Chemists. The methods for animal foods were those used by Atwater and Woods.[1]

Descriptions of specimens and results of the analyses are given in the following pages. Table 1 gives the composition of the food materials as found in the market, including both edible portion and refuse. Table 2 shows the composition of the edible portion calculated on the basis of the water content at the time the samples were taken. Table 3 gives the composition of the water-free substance of the edible portion. In addition to the ordinary data of composition, Tables 2 and 3 also give the estimated fuel value or potential energy of the foods. These estimates are made by assuming each gram of protein or carbohydrates to furnish 4.1 calories of energy and each gram of fat 9.3 calories.

DESCRIPTION OF SAMPLES OF FOOD MATERIALS.[2]

63. *Beef, sirloin steak.*—From native (Alabama) beef. Purchased in Auburn at 10 cents per pound. Contains more fat than average samples obtainable in this market.

180. *Beef, rib roast.*—From native beef, quite fat. Purchased in Auburn at 10 cents per pound.

226. *Beef, roast, cross ribs.*—From near shoulder above the brisket. Obtained in Birmingham, Ala.

231. *Beef, round steak.*—From native beef. Purchased at Tuskegee at 10 cents per pound.

255. *Beef, round steak.*—From native beef. Sample quite fat. Purchased in Auburn at 10 cents per pound.

256. *Beef, round steak.*—From native beef. Much leaner than the preceding.

333. *Beef, shoulder steak.*—From native beef. Purchased in Tuskegee at 10 cents per pound. Very deficient in fat.

334. *Beef, shoulder steak.*—From native beef. Purchased in Tuskegee at 10 cents per pound. Very deficient in fat.

336. *Beef, shoulder steak.*—From native beef. Purchased in Auburn at 10 cents per pound.

351. *Beef, shoulder roast.*—From native beef. Purchased in Auburn at 10 cents per pound.

1589. *Mutton, shoulder.*—Obtained from a packing establishment in Birmingham, Ala. Quite fat; quality above the average of mutton sold in Alabama markets.

Connecticut Storrs Sta. Rpt. 1891, p. 17; U. S. Dept. Agr., Office of Experiment Stations Bul. 29.

The numbers used in this list are as a rule those employed in an unpublished compilation of analyses of American food materials.

2028. *Pork, ribs.*—Side of fresh pork, including ribs. Very fat. Animal medium size. Purchased in Auburn at 10 cents per pound.

2069. *Pork, smoked ham.*—From packing house in Birmingham. Contains more fat than most imported hams.

2071. *Pork, smoked ham.*—Home cured ham from native Alabama hog. Cured in smokehouse and also salted. Sample somewhat fatter than ordinary home-cured hams.

2091. *Pork, smoked shoulder.*—From packing establishment in Birmingham. From animal of medium size.

2125. *Pork, salt sides.*—Cured and salted. Purchased at a store in Auburn at 8 cents per pound. Sample contained three short ribs.

2126. *Pork, salt sides.*—Cured and salted. Purchased in Auburn at 8 cents per pound. Sample contained one rib. Proportion of fat somewhat below the average of salt pork on the market.

2127. *Pork, salt sides.*—Side of pork cured as above. Purchased in Auburn at 8 cents per pound. This and the two preceding samples were packed in Western markets.

2128. *Pork, salt sides.*—Cured salt pork or bacon. Obtained from packing establishment in Birmingham. Not so strongly salted as most of the Western meats.

2129. *Pork, salt sides.*—Home-cured bacon cured in old-style smokehouse and also salted. Sample quite fat.

2130. *Pork, salt sides.*—Purchased in Tuskegee.

2702. *Chicken.*—Full grown. Purchased in Auburn. Price 25 cents.

2705. *Chicken.*—Full grown.—Larger than average. Price 30 cents.

2756. *Eggs.*—Purchased in Auburn at 12½ cents per dozen.

4042. *Lard.*—Purchased as pure leaf lard in Auburn.

4049. *Lard.*—Purchased in Tuskegee.

[1]11. *Butter.*—Composite sample from churnings of three families in vicinity of Tuskegee.

[1]60. *Butter.*—Purchased in Auburn.

5036. *Corn meal.*—Unbolted. From Tuskegee. Proportion of bran approximately normal.

5037. *Corn meal.*—Unbolted. From Tuskegee. Price 75 cents per bushel. Proportion of bran quite high.

5038. *Corn meal.*—Unbolted. From Tuskegee. Purchased from the Institute.

5039. *Corn meal.*—Purchased from store on plantation near Tuskegee. Appears to have been partially bolted.

5040. *Corn meal.*—Quality about the same as preceding.

5041, 5042. *Corn meal.*—Native. Unbolted.

5343. *Wheat flour.*—From Tuskegee.

5344. *Wheat flour.*—From Tuskegee. Practically all flours found in Alabama are made from Western wheat in Western mills.

6026. *Molasses.*—A dark-colored centrifugal molasses from Louisiana.

---

[1] These are the Alabama Experiment Station laboratory numbers.

It represents fairly well the quality of the average molasses consumed on the plantations. During the fall and early winter, however, home-made cane sirup is largely used.

6030. *Molasses.*—Native.

6031. *Sorghum.*

6528. *Beans, string.*—Ordinary running variety. This particular variety is more commonly cultivated than any other during the latter part of the season.

6529. *Beans, butter.*—Common variety. Purchased in pod.

6547. *Beets.*—Ordinary blood-red variety. Small size.

6557. *Cabbage.*—An early variety. Head of medium size.

6560. *Collards.*—This is a variety consumed to a large extent by the colored population, although the consumption of this vegetable is greater at some seasons than it was during the study.

6586. *Corn, green.*—Ordinary garden variety. Purchased in the ear.

6651. *Cowpeas, green.*—This is the ordinary cowpea, fresh from the vine, and represents quite an early variety. The peas as purchased had been removed from the pod, and hence the refuse is not taken into consideration.

6642. *Cowpeas, dried.*—Grown on Institute farm, Tuskegee. This sample is the ordinary speckled pea and is more generally grown than any other variety. It is planted in the early summer and harvested in latter part of summer or early fall. The pea fresh from the pod is an important article of food in the early fall, and the dried pea, when properly cooked, becomes quite soft and palatable. It is little used out of its immediate season, however.

6643. *Cowpeas, dried.*—Variety known as the Clay pea.

6644. *Cowpeas, dried.*—Reddish brown in color. Not so much used as the speckled pea.

6645. *Cowpeas, dried.*—Speckled variety.

6646. *Cowpeas, dried.*—Black variety. Not much used in this section. The four last samples were obtained in the Auburn market.

6652, 6653. *Cowpeas, dried.*—From Tuskegee.

6590. *Cucumbers.*—Large size. Form very much elongated.

6595. *Turnip greens (salad).*—This sample was gathered with a view to the use of the tops for the preparation of the so-called turnip salad, while sample 6938 was selected with a view to the use of the root or the turnip proper.

6609. *Okra.*—Also an early variety. Pods somewhat elongated.

6743. *Potatoes.*—A late variety. Small size.

6896. *Squash.*—Ordinary crook-neck variety, which is more commonly grown in this section than any other variety. Medium size. Average weight about half a pound.

6860, 6861. *Sweet potatoes.*—Ordinary white. Compact form. As these samples were procured for analysis out of season, and as the specimens contained, as a consequence, less than the normal amount of water, it was deemed best in giving the analysis of the fresh material to calculate the same on a basis of 64.32 per cent water, this being the average water content of 15 samples of potatoes previously analyzed in proper season.

6918. *Tomatoes.*—A large, early variety. The only variety obtainable at this early date in the season.

6938. *Turnips.*—A small variety, with very long, tapering root.

8146. *Watermelon.*—Large rounded variety. The melons selected for analysis had an average weight of 26 pounds.

Professor Ross, in submitting his report of the foods analyzed, writes as follows:

We have, in these investigations, analyzed specimens of all the more important meats in use in this section. The vegetable products selected for analysis were those obtainable during the latter part of the spring and early part of the summer. They represent quite fairly the vegetable foods in use throughout a very considerable area of country, and their composition, as shown by analysis, indicates that a number of the more common food materials of this class are undoubtedly of a high nutritive value.

While practically all of the vegetables included in our investigations are cultivated and utilized by the white portion of the agricultural population, the variety and number of vegetable foods grown and utilized by the negro laborer is much less comprehensive. A personal inspection of a number of gardens cultivated by the colored population in this vicinity reveals the fact that not more than half a dozen varieties of vegetable foods are grown by those producing vegetables for their own consumption, and in a number of cases the variety is much smaller.

The vegetables most generally cultivated are turnips, collards, string beans, corn, and Irish potatoes, while cabbage is grown to a somewhat less extent. At a somewhat later period in the season cowpeas replace some of the vegetables named, while in the latter part of the summer or in the early fall the sweet potato comes into use.

The turnips are cultivated almost exclusively for the tops or "turnip greens," which are used as a pot herb,[1] which is locally called "salad," while collards are utilized in season in similar manner. These two vegetables are probably employed for food purposes by the colored laborer to a greater extent than almost any other vegetable food, and either one or the other can be obtained at almost any time from April to October, as turnips are grown from both spring and fall plantings, while collards can be obtained at almost any time during the middle and latter part of this period. The importance of these facts will be observed when a reference is made to the composition of both collards and turnip greens, the percentages of protein, as calculated on a water-free basis, being in excess of that of cowpeas. As the ordinary diet of the average colored laborer is characterized by the presence of excessive quantities of fat and carbohydrates and corresponding deficiencies of protein, it is of course quite obvious that the employment of the above vegetable foods will serve to some extent at least to overcome the great disproportion between fuel ingredients and flesh-forming constituents which exists in the dietary of the average laborer.

--- --- --

[1] That is, they are cooked before being eaten.

TABLE 1.—*Composition of Alabama food materials as purchased (including both edible portion and refuse).*

Kind of food material.	Reference number.[1]	Refuse.	Water.	Protein.	Fat.	Carbo-hydrates.	Ash.	Fuel value per pound.
		Per ct.	*Per ct.*	*Per cent.*	*Per ct.*	*Per cent.*	*Per ct.*	*Calories.*
ANIMAL FOOD.								
Beef								
Sirloin steak	63	12.0	59.4	17.5	10.1	....	1.0	750
Rib	180	26.7	50.3	12.4	10.0	....	.6	650
Cross rib	226	12.8	57.4	16.1	13.0	....	.7	850
Round	231	3.4	72.8	21.4	1.3	....	1.1	155
Do	255	4.8	66.7	18.1	9.4	....	1.0	735
Do	256	6.5	68.8	18.7	4.8	....	1.2	550
Average	......	4.9	69.4	19.4	5.2	....	1.1	580
Shoulder steak	333	12.5	65.8	19.6	1.1	....	1.0	410
Do	334	17.1	62.3	18.4	1.2	....	1.0	390
Do	336	7.3	68.8	18.5	4.3	....	1.1	525
Do	351	16.1	62.3	14.5	6.0	....	.8	525
Average	......	13.4	64.8	17.7	3.1	....	1.0	460
Mutton, shoulder	1589	14.0	55.7	15.5	13.4	....	.8	855
Pork:								
Ribs	2028	12.0	34.9	10.6	41.9	....	.6	1,965
Smoked ham[2]	2069	18.8	35.9	11.8	29.8	....	3.7	1,480
Do	2071	2.0	22.0	14.0	55.6	....	6.4	2,605
Average	......	10.4	29.0	12.9	42.7	....	5.0	2,040
Smoked shoulder	2091	17.7	40.8	13.3	29.7	....	4.5	1,250
Salt sides[2]	2125	9.1	13.3	6.2	67.0	....	4.4	2,940
Do[2]	2126	9.5	20.9	9.4	53.0	....	7.2	2,410
Do[2]	2127	2.9	14.6	7.9	68.6	....	6.0	3,040
Do	2128	7.2	16.7	8.6	64.3	....	3.2	2,875
Do	2129	8.7	7.1	7.0	72.8	....	4.4	3,200
Do	2130	9.5	10.8	6.0	70.9	....	2.8	3,105
Average	......	7.8	13.9	7.5	66.1	....	4.7	2,930
Poultry:								
Chicken	2702	31.4	52.4	14.4	1.1	....	.7	315
Do	2705	23.3	49.1	14.8	12.0	....	.8	780
Eggs	2756	9.9	66.8	13.2	9.3	....	.8	640
Lard[2]	4042	......	.4	.3	99.3	....	......	4,195
Do[2]	4049	......	......	.2	99.8	....	......	4,215
Butter	[3]14	......	15.8	1.2	82.4	....	.6	3,500
Do	[3]60	......	23.6	1.3	70.3	....	4.8	2,990
Average	......	......	19.7	1.2	76.4	....	2.7	3,245
VEGETABLE FOOD.								
Corn meal, unbolted	5036	10.1	10.4	7.3	4.1	66.8	1.3	1,550
Do	5037	24.1	9.2	6.5	3.5	55.7	1.0	1,305
Do	5038	8.9	10.5	7.8	4.3	67.3	1.2	1,580
Do	5039	4.2	10.4	7.5	4.4	72.2	1.3	1,670
Do	5040	5.1	10.4	7.5	4.3	71.4	1.3	1,650
Do	5041	11.3	10.2	7.7	4.0	65.6	1.2	1,535
Do	5042	13.0	10.8	8.0	4.5	62.6	1.1	1,505
Average	......	10.9	10.3	7.5	4.1	66.0	1.2	1,540
Wheat flour[2]	5343	......	11.1	9.9	.9	77.7	.4	1,605
Do[2]	5344	......	10.7	9.3	.8	78.9	.3	1,675
Average	......	......	10.9	9.6	.9	78.3	.3	1,670
Molasses, New Orleans[2]	6026	......	23.1	1.3	.1	68.3	7.2	1,300
Molasses, native	6030	......	26.2	.9	.2	72.1	.6	1,365
Molasses, sorghum	6031	......	22.9	.7	.1	75.1	1.2	1,415
Beans, string	6528	7.0	82.6	2.1	.2	7.3	.8	185
Beans, butter	6529	50.5	29.1	4.7	.3	14.4	1.0	370
Beets	6547	16.6	70.9	.9	.1	10.9	.6	225
Cabbage	6557	19.4	74.5	1.2	.1	4.2	.6	105
Collards	6560	55.3	39.5	1.5	.2	2.9	.6	90
Corn, green	6586	61.1	28.1	1.4	.4	8.7	.3	205

[1] The numbers used in an unpublished compilation of analyses of American food materials.
[2] Not native Alabama food materials.
[3] Alabama Experiment Station laboratory numbers.

TABLE 1.—*Composition of Alabama food materials as purchased (including both edible portion and refuse)*—Continued.

Kind of food material.	Reference number.	Refuse.	Water.	Protein.	Fat.	Carbohydrates.	Ash.	Fuel value per pound.
VEGETABLE FOOD—cont'd.		*Per ct.*	*Per ct.*	*Per cent.*	*Per ct.*	*Per cent.*	*Per ct.*	*Calories.*
Cowpeas, dried...	6642	....	11.9	19.9	1.5	62.9	3.8	1,605
Do...	6643		11.2	21.8	1.2	62.5	3.3	1,620
Do...	6644		11.6	22.1	1.3	61.4	3.6	1,610
Do...	6645		12.2	22.1	1.4	60.7	3.6	1,600
Do...	6646		11.1	22.6	1.4	61.1	3.8	1,615
Do...	6652		11.9	22.3	1.1	61.3	3.4	1,600
Do...	6653		14.6	20.8	1.6	59.3	3.7	1,555
Average...			12.1	21.7	1.3	61.3	3.6	1,600
Cowpeas, green...	6654	......	65.9	9.4	.6	22.7	1.4	620
Cucumbers...	6590	32.6	63.9	.6		2.7	.2	60
Greens, turnip tops...	6595		65.9	9.5	.6	22.6	1.4	620
Okra...	6609	13.7	80.2	1.0	.1	4.6	.4	110
Potatoes...	6743		79.9	1.9		17.1	1.1	355
Squash...	6898	26.9	68.7	.6	.1	3.3	.4	75
Sweet potatoes...	6800		64.3	1.6	.3	32.8	1.0	650
Do...	6861		64.3	2.1	.9	31.4	1.3	660
Average...			64.3	1.9	.6	32.1	1.1	655
Tomatoes...	6918	2.0	90.8	1.4	.3	4.7	.8	125
Turnips...	6938	31.6	61.2	1.0	.1	6.3	.4	140
Watermelon...	8146	60.9	36.0	.2	.1	2.7	.1	60

TABLE 2.—*Composition of fresh, edible portion of Alabama food materials.*

Kind of food material.	Reference number.[1]	Water.	Protein.	Fat.	Carbohydrates.	Ash.	Fuel value per pound.
ANIMAL FOOD.							
Beef.		*Per cent.*	*Per cent.*	*Per cent.*	*Per cent.*	*Per cent.*	*Calories.*
Sirloin steak...	63	67.5	19.9	11.5	....	1.1	855
Rib...	180	68.6	16.9	13.6	....	.9	890
Cross rib...	226	65.8	18.4	14.9	....	.9	970
Round...	241	75.4	22.1	1.3	....	1.2	465
Do...	255	70.0	19.0	9.9	....	1.1	770
Do...	256	73.6	20.0	5.1	....	1.3	585
Average...		73.0	20.4	5.4	....	1.2	605
Shoulder steak...	333	75.2	22.4	1.3	....	1.1	470
Do...	334	75.1	22.3	1.4	....	1.2	475
Do...	336	74.2	20.0	4.7	....	1.1	570
Do...	351	74.5	17.4	7.1	....	1.0	625
Average...		74.8	20.5	3.6	....	1.1	535
Mutton, shoulder...	1589	65.2	18.2	15.6	....	1.0	995
Pork:							
Rib...	2028	39.7	12.0	47.7	....	.6	2,235
Smoked ham[2]...	2069	44.2	14.5	36.7	....	4.6	1,820
Do...	2071	22.4	14.3	56.8	....	6.5	2,665
Average...		33.3	14.4	46.7	....	5.6	2,240
Smoked shoulder...	2091	49.6	16.1	28.8	....	5.5	1,515
Salt sides[2]...	2125	14.7	6.8	73.7	....	4.8	3,235
Do[2]...	2126	23.1	10.4	58.6	....	7.9	2,665
Do[2]...	2127	15.1	8.2	70.6	....	6.1	3,130
Do...	2128	18.0	9.3	69.3	....	3.4	3,100
Do...	2129	7.7	7.7	79.7	....	4.9	3,505
Do...	2130	12.0	6.6	78.3	....	3.1	3,425
Average...		15.1	8.2	71.7	....	5.0	3,180
Poultry:							
Chicken...	2702	76.3	21.1	1.6	....	1.0	460
Do...	2705	64.0	19.1	15.6	....	1.0	1,020
Eggs...	2756	74.1	14.7	10.3	....	.9	710
Lard[2]...	4042	.4		.3	99.3	....	4,195
Do[2]...	4049	....	.2	99.8	....	....	4,215

[1] The numbers used in an unpublished compilation of analyses of American food materials.
[2] Not native Alabama food material.

TABLE 2.—*Composition of fresh, edible portion of Alabama food materials—* Continued.

Kind of food material.	Reference number.	Water.	Protein.	Fat.	Carbo-hydrates.	Ash.	Fuel value per pound.
ANIMAL FOOD—continued.							
		Per cent.	*Per cent.*	*Per cent.*	*Per cent.*	*Per cent.*	*Calories.*
Butter	[1]14	15.8	1.2	82.4		0.6	3,500
Do	[1]60	23.6	1.3	70.3		4.8	2,990
Average		19.7	1.3	76.3		2.7	3,245
VEGETABLE FOOD.							
Corn meal, unbolted	5036	11.6	8.1	4.6	74.3	1.4	1,725
Do	5037	12.1	8.6	4.6	73.4	1.3	1.720
Do	5038	11.5	8.5	4.7	73.9	1.4	1.730
Do	5039	10.9	7.8	4.6	75.4	1.3	1,740
Do	5040	11.0	7.9	4.5	75.3	1.3	1,740
Do	5041	11.5	8.7	4.6	73.9	1.3	1.730
Do	5042	12.4	9.3	5.2	71.9	1.2	1.730
Average		11.6	8.4	4.7	74.0	1.3	1.730
Wheat flour [2]	5343	11.1	9.9	.9	77.7	.4	1,665
Do [2]	5344	10.7	9.3	.8	78.9	.3	1.675
Average		10.9	9.6	.9	78.3	.3	1 670
Molasses, New Orleans [2]	6026	23.1	1.3	.1	68.3	7.2	1,300
Molasses, native	6030	26.2	.9	.2	72.1	.6	1,365
Molasses, sorghum	6031	22.9	.7	.1	75.1	1.2	1 415
Beans, string	6528	88.7	2.3	.2	7.9	.9	215
Beans, butter	6529	58.9	9.4	.6	29.1	2.0	740
Beets	6547	85.0	1.1	.1	13.1	.7	270
Cabbage	6557	92.5	1.5	.1	5.2	.7	130
Collards	6560	88.3	3.3	.5	6.5	1.4	205
Corn, green	6586	72.1	3.7	1.1	22.4	.7	530
Cowpeas, dried	6612	11.9	19.9	1.5	62.9	3.8	1.605
Do	6643	11.2	21.8	1.2	62.5	3.3	1.620
Do	6644	11.6	22.1	1.3	61.4	3.6	1 610
Do	6645	12.2	22.1	1.4	60.7	3.6	1,600
Do	6646	11.1	22.6	1.4	61.1	3.8	1.615
Do	6652	11.9	22.3	1.1	61.3	3.4	1.600
Do	6653	14.6	20.8	1.6	59.3	3.7	1,555
Average		12.1	21.7	1.3	61.3	3.6	1,600
Cowpeas, green	6654	65.9	9.4	.6	22.7	1.4	620
Cucumbers	6590	94.7	.9	.1	4.0	.3	95
Greens, turnip tops	6595	65.9	9.5	.6	22.6	1.4	620
Okra	6609	92.9	1.2	.1	5.3	.5	125
Potatoes	6743	79.9	1.9		17.1	1.1	355
Squash	6896	93.9	.9	.2	4.4	.6	105
Sweet potatoes	6860	64.3	1.6	.3	32.8	1.0	650
Do	6861	64.3	2.1	.9	31.4	1.3	660
Average		64.3	1.9	.6	32.1	1.1	655
Tomatoes	6918	92.6	1.5	.3	4.8	.8	130
Turnips	6928	88.7	1.5	.1	9.2	.5	205
Watermelon	8146	92.0	.6	.2	6.9	.3	150

[1] Alabama Experiment Station laboratory numbers.
[2] Not native Alabama food material.

TABLE 3.—*Composition of water-free substance of Alabama food materials.*

Kind of food material.	Reference number.[1]	Nitrogen.	Protein.	Fat.	Carbo-hydrates.	Ash.
ANIMAL FOOD.						
Beef:		*Per cent.*	*Per cent.*	*Per cent.*	*Per cent.*	*Per cent.*
Sirloin steak	63	9.41	61.4	35.3		3.3
Rib	180	8.41	53.7	43.4		2.9
Cross rib	226	8.40	53.8	43.7		2.5
Round	231	13.69	89.8	5.4		4.8
Do	255	10.63	63.3	33.1		3.6
Do	256	12.13	75.8	19.4		4.8
Average		11.95	76.3	19.3		4.4

[1] The numbers used in an unpublished compilation of analyses of American food materials.

TABLE 3.—*Composition of water-free substance of Alabama food materials*—Continued.

Kind of food material.	Reference number.	Nitrogen.	Protein.	Fat.	Carbohydrates.	Ash.
ANIMAL FOOD—continued.						
Beef—Continued.		*Per cent.*	*Per cent.*	*Per cent.*	*Per cent.*	*Per cent.*
Shoulder steak...	333	13.84	90.3	5.2	....	4.4
Do...	334	13.89	89.6	5.6	....	4.8
Do...	336	11.94	77.6	18.0	....	4.4
Do...	351	10.92	68.0	28.0	....	4.0
Average...	....	12.65	81.4	14.2	....	4.4
Mutton, shoulder...	1589	8.31	52.2	44.9	....	2.9
Pork:						
Ribs...	2028	3.25	19.9	79.0	....	1.1
Smoked ham[1]...	2069	3.74	26.0	65.8	....	8.3
Do...	2071	2.83	18.4	73.2	....	8.4
Average...	....	3.28	22.2	69.5	....	8.3
Smoked shoulder...	2091	4.50	31.9	57.2	....	10.9
Salt sides[1]...	2125	1.31	7.9	86.4	....	5.7
Do[1]...	2126	2.31	13.6	76.1	....	10.3
Do[1]...	2127	1.54	9.6	83.2	....	7.2
Do...	2128	1.98	11.3	84.5	....	4.2
Do...	2129	1.08	8.3	86.4	....	5.3
Do...	2130	....	7.5	89.0	....	3.5
Average...	....	....	12.9	80.4	....	6.7
Poultry:						
Chicken...	2702	13.60	89.1	6.7	....	4.2
Do...	2705	8.58	53.8	43.5	....	2.7
Eggs...	2756	7.97	56.9	39.7	....	3.4
Lard[1]...	4042	.04	.3	99.7	....	...
Do[1]...	4049	....	.2	99.8	....	...
Butter...	711	.22	1.1	97.8	....	.8
Do...	709	....	1.6	92.1	....	6.3
Average...	....	....	1.5	95.0	....	3.5
VEGETABLE FOOD.						
Corn meal, unbolted...	5036	....	9.1	5.2	84.1	1.6
Do...	5037	....	9.8	5.2	83.5	1.5
Do...	5038	....	9.6	5.3	83.6	1.5
Do...	5039	....	8.7	5.2	84.6	1.5
Do...	5040	....	8.9	5.1	84.5	1.5
Do...	5041	....	9.9	5.1	83.5	1.5
Do...	5042	....	10.5	6.0	82.1	1.4
Average...	....	....	9.5	5.3	83.7	1.5
Wheat flour[1]...	5343	....	11.2	1.0	87.4	.4
Do[1]...	5344	....	10.3	.9	88.4	.4
Average...	....	....	10.7	1.0	87.9	.3
Molasses, New Orleans[1]...	6026	....	1.7	.1	88.9	9.4
Molasses, native...	6030	....	1.2	.2	97.8	.8
Molasses, sorghum...	6031	....	.9	.1	97.4	1.6
Beans, string...	6528	....	20.2	1.9	70.2	7.7
Beans, butter...	6529	....	23.0	1.4	70.7	4.9
Beets...	6547	....	7.4	.3	87.3	5.0
Cabbage...	6557	....	19.4	1.5	69.7	9.4
Collards...	6560	....	28.1	4.1	55.7	12.1
Corn, green...	6586	....	13.3	3.9	80.2	2.6
Cowpeas, dried...	6642	....	22.5	1.7	71.4	4.4
Do...	6643	....	24.5	1.4	70.4	3.7
Do...	6644	....	25.0	1.5	69.4	4.1
Do...	6645	....	25.2	1.6	69.1	4.1
Do...	6646	....	25.4	1.6	68.7	4.3
Do...	6652	....	25.3	1.2	69.6	3.9
Do...	6653	....	24.4	1.8	69.1	4.4
Average...	6654	....	24.6	1.6	69.7	4.1
Cowpeas, green...	6654	....	27.7	1.8	66.4	4.1
Cucumbers...	6590	....	17.2	.8	75.8	6.2
Greens, turnip tops...	6595	....	29.4	4.5	19.6	16.5
Okra...	6609	....	16.9	1.1	75.4	6.6
Potatoes...	6713	....	9.4	.2	84.9	5.5
Squash...	6896	....	14.5	3.2	72.6	9.7
Sweet potatoes...	6860	....	4.5	.9	91.8	2.8
Do...	6861	....	5.9	2.4	88.0	3.7
Average...	....	....	5.2	1.7	89.9	3.2
Tomatoes...	6918	....	19.6	4.3	65.4	10.7
Turnips...	6958	....	63.2	.9	81.1	4.8
Watermelon...	8146	....	7.3	2.9	86.6	3.2

[1] Not native Alabama food material. [2] Alabama Experiment Station laboratory numbers.

THE INVESTIGATIONS AT TUSKEGEE.

THE REGION AND THE PEOPLE.

The region around Tuskegee and the colored people who make up the larger part of its population and their ways of living are described as follows by Mr. H. M. Smith:

The region.—Tuskegee is situated in the eastern part of Alabama on the edge of the so-called black belt. The term black belt is applied to a region with boundaries not very sharply defined, but extending from the Gulf of Mexico northward as far as central or northern Alabama and Georgia, and westward to Louisiana and Texas. Two explanations are given of the term "black belt," one ascribing it to the soil, which in a large part of the region is dark in color; the other to the preponderance of the colored population. Either explanation would fit the case. The naturally fertile soil made slave labor profitable before the war. The negro population was then, and still continues to be, large, so that to-day in the county in which Tuskegee is located (Macon), the ratio of the negroes to the whites is over three to one.

The negroes about Tuskegee.—The negroes of this section, in which there are but few large towns, are mostly engaged in farming. Very few as yet own any land; the larger number work small farms rented from white proprietors. As a class they are improvident, they have very little ambition, and little incentive to work because of their ignorance of any better conditions of living than those immediately around them. Their wants like their resources are few, so that with all their poverty they appear to be a happy and contented people. In the neighborhood of the Tuskegee Institute were, however, most noteworthy indications of progress. Comparatively few of the families lived in one-room log cabins; a number had frame houses with several rooms, respectable furniture, and more or less of the conveniences of modern life such as are found in the houses of the working classes of other regions.

Negro cabins.—In the country practically all the negroes live in cabins, generally built of logs, with only one, or at most two rooms. The spaces between the logs were either left open, admitting free passage of the wind in winter as well as in summer, or were chinked with earth or occasionally with pieces of board. The roofs were covered with coarse shingles or boards and were apt to be far from tight. The windows had no sash or glass, but instead, wooden blinds, which were kept open in all weather to admit the light. The cabins generally stood on posts a few feet from the ground; the door was approached by a box or a few broken steps. The open space between the floor of cabin and the ground was generally occupied by dogs, with which but few families were not supplied.

The one-room cabins had a door in front, a fireplace on one side, and perhaps one or two windows in the side or rear walls. The one room served for kitchen and living and sleeping room. Occasionally a small annex was built on the rear and served as a storeroom.

The two-room cabins differed from those of one room in that they were longer and divided in three parts. The middle division had a roof and floor but was open front and back, and thus served as a sort of porch. One of the rooms served as kitchen and living room, the other as sleeping room. The chimneys were built of small logs laid against the wall on the outside and reached no higher than the roof, and in some cases not quite so high. These logs were chinked with clay, with which the interior was also lined. The chimney opened into the cabin, making a fireplace about four feet square. The cabin floors were made of rough boards with cracks of varying widths. In one cabin situated close by a swamp, which abounded in moccasins, some of the cracks in the floor were an inch or more wide. In response to a question put in all serious-ness whether the snakes could not crawl in through them, the woman replied, "Oh, yes, they gets in sometimes, but I bresh 'em out."

The furniture of these cabins was very limited, consisting of one or two roped bedsteads with corn-shuck mattresses and patchwork quilts, a small portable wooden cupboard containing a few dishes, a wooden chest or old trunk used as a receptacle for both food and clothing, a cheap pine table, a few homemade chairs, a pair of andirons and an iron pot in the fireplace, an earthenware jar used for a churn, and sometimes a clock. Occasionally there would be a few books or a picture.

In addition to the house there was generally a small, rough shed or barn with walls on either three or four sides. Both the cabins and sheds or barns were as a rule much dilapidated.

Gardens and farms.—A few of the families had gardens: that is to say, small patches of ground close to the houses were used to grow collards, turnips, and occasionally some other vegetables.

The farms occupied by individual tenants varied from 20 to 60 acres. They were commonly spoken of as one, two, or three mule farms. The area which could be cultivated by one mule was variously stated at from 25 to 40 acres.

Of the field crops used for food the most common were corn, sweet potatoes, sugar cane, and sorghum, the last two being used to make molasses for home consumption. On only one of the plantation farms visited were any cowpeas or peanuts grown. Quite a number raised corn for the nourishment of their families and live stock, though very few raised enough to supply all their needs. Some were so improvi-dent as to sell it as soon as it was marketable, even though they had to buy it back again later in the season. The staple crop was cotton, to which the larger part of the cultivated land was devoted. The status of a negro farmer here is decided mainly by the number of bales of cot-ton he can produce in a year. A greater diversity of crops is one of the great needs. Tuskegee Institute is making an effort to encourage the cultivation of corn, peas. and other crops, both for food of man and to aid in the keeping of more and better live stock.

Live stock.—The live stock of the negro farmers varied with the season. Generally each one had a mule or an ox, one or more pigs, frequently a cow, and very often hens. The oxen were poor and underfed. It was interesting to compare the small loads drawn by these small, half-starved creatures with those drawn by the cattle belonging to the Institute, which were of good breeds and thoroughly well fed and cared for.

The negro farmer's life.—The negro farmer generally works about seven and a half months during the year. The busy seasons are two, that of planting and cultivating the cotton, which begins in March and lasts until the end of June, and that of cotton picking. which begins about the middle of August and continues until the latter part of November. The rest of the time is devoted to visiting. social life, revivals or other religious exercises, and to absolute idleness. The working season opens with the plowing of the land: planting begins in April, and is followed by the "chopping." i. e., the hoeing and thinning out of the cotton. This is done by both men and women. They swing their large hoes in a slow, regular movement. frequently keeping time to the tune of some plantation song. After the chopping and while the crop is maturing, is a period of rest called "laying-by time." This period the negro enjoys by holding " bush meetings " (camp meetings) and visiting among his friends on neighboring plantations. Whole families thus visit with each other for a week at a time. By the middle of August the cotton begins to open, and then comes a busy season of picking. when men and women, old and young. even the little children, are pressed into the work. During this season, as well as that of planting and chopping. work in the field begins at sunrise and lasts until sunset. with a short rest at noon. Toward the end of November the picking is over: then comes a season of general festivities. During the winter months the men pass much of their time in the house by the fire. One more thrifty than his neighbor may perhaps spend a few days fixing up his fences. or making chairs and baskets. But few of the negro farmers work on Saturday even in the busy season: instead. the whole family goes to the village and does its marketing, which consists principally of the purchase of a little corn meal and salt pork for the next week's rations.

The negro's earnings and business methods.—At the end of the season when the farmer has sold his cotton he has. if the crop was good. a little money, but this is usually soon spent and the rest of the year he lives from hand to mouth. He may occasionally collect a small load of dead limbs and fragments of stumps of trees and sell them for what he can get. He also earns a little money in other ways. The rate of wages may be inferred from the fact that when he works out his road tax, or that of some white man who employs him for the purpose. he is allowed from 40 to 50 cents per day. When the negro has money he is ready to spend it for almost anything, and the skillful trader may urge goods upon him the purchase of which is most extravagant. After his cotton is sold and the mortgage on the crop is paid, he may spend a large part of the balance for a sewing machine or a modern cooking range, which are ultimately returned to the dealer at a large sacrifice. When he has no money he will buy on credit as much and as long as he can.

The mortgage system.—The negro farmers in this region have a custom of mortgaging their crops, which comes partly from necessity but largely from improvidence. A tenant is very apt to be without the money needed at the beginning of the season to buy seed, tools, or a mule, with which to commence work, and later for the food and clothing necessary for himself and family until his crop ripens. He tries to get over the difficulty by signing a "waive note," giving the first right to so much of the crop as may be necessary to cover the indebtedness incurred to meet these needs. The person to whom this mortgage is given may be the owner of the land he tills, or the proprietor of the store at which his supplies are purchased. The negro knows but little of accounts, and the white man who holds the mortgage keeps them; the rates of interest are high, and the mortgagee is not always generous or even just. At the end of the season if the crop is a failure the debtor has absolutely nothing; if the crop and the creditor's accounts are favorable there may be a fair balance on the debtor's side. This evil is often charged in some sections to the extortion and injustice of the white man: but it seems probable that the shiftlessness and improvidence of the negro which inevitably accompany his ignorance are largely to blame. The cure will only come with education; this must be industrial as well as intellectual. The influence of such an institution as that at Tuskegee in this direction is most salutary and fortunate.

Food of the negro.—The staple foods of the negroes of this region are fat salt pork, corn meal, and molasses. Of late, since wheat flour has become so cheap, it has been considerably used. The molasses is made from sorghum, or "millet," as it is called in this region, and sugar cane, both of which are grown in considerable quantities. The molasses from sorghum is generally preferred to that from cane. The molasses is made on the farms by a very primitive process. This consists in passing the cane between rollers to squeeze out the juice, and boiling the latter in open pans, which are set on furnaces roughly built of stone and clay. There are persons who go about from farm to farm with the rollers and make the molasses. Individual farmers who have no conveniences for making sirup carry their cane to other farms where it is worked. Only a part of the molasses used by the farmer is made on the farms, the rest is bought at the stores with other commodities.

Part of the corn meal is made from the corn grown on the farms, the rest is bought from dealers, and is uniformly unbolted.

The pork consists mostly of the fat sides, butchered and salted in the meat-packing houses of Chicago and elsewhere, and brought in large quantities to the Southern market. Some pork is produced on the farms, but comparatively few swine were seen on those visited, nor was any kind of meat but fat pork, not even ham or shoulder, seen in any of the farmhouses. In the home of a well-to-do carpenter, which is located near the Institute (p. 23), fresh beef and mutton were used during the two weeks of a dietary study. Probably this case was exceptional; indeed, the only kind of meat which seemed to be in at all common use among the country people was fat pork. Whenever they spoke of meat they always meant fat pork. Some of them knew it by no other name, nor did they seem to know much of any other meat except that of opossum and rabbits, which they occasionally hunted, and of chickens which they raised to a limited extent.

Even among the white population in the village of Tuskegee the use of fresh meat was not at all large. The table of the hotel was well supplied with fried ham and pork, but there was comparatively little beef. Fresh beef was to be had at the market on two or three days in the week. This limited use of fresh meats could not be attributed to any lack of generous diet, for the tables of white people were bountifully spread. It seemed due to the agricultural conditions which obtain in the region, and to the difficulty of keeping fresh meat in the warm climate. The climate is not favorable to the growth of ordinary grasses which are so abundant in the beef-producing regions; comparatively few cattle are raised, and the meat is less fat, and less tender and juicy than that from the grazing regions farther north. In Knoxville, Tenn., for instance, where dietary studies have been lately made,[1] in a region where grass and corn are abundant, the native beef was much more plentiful and appetizing, and the specimens analyzed in connection with dietary studies were considerably fatter, than those from Tuskegee and elsewhere in Alabama, and were in this respect more like those of meats in the Northern markets. Veal and mutton are even less common than beef. No sheep were seen in the country about Tuskegee and there are very few in the region.

The scarcity of fresh meat and the difficulty of preserving it doubtless goes far toward explaining the dietary tastes and habits of the people in general in this region, if not elsewhere in the South. The managers of the colored schools find their students decidedly averse to a diet materially different from that of salt pork, corn meal, and molasses, to which they have been accustomed at home.

The colored families near the village of Tuskegee, and some in the country, kept cows and had milk and butter. For making butter they used small, dash churns of glazed earthenware called "splashers," which are usually about 15 inches high by 8 in diameter. The fresh milk was put directly into the churn and successive milkings were added until it contained from $1\frac{1}{2}$ to 2 gallons, and the whole churned without any attempt at removing the cream. The churning was done about once in two days, and from the above amount of milk a small saucerful of a soft, white, and watery butter would be obtained. The people made no attempt at working it, nor did they add salt, but ate it fresh. The buttermilk was drunk with decided relish.

No cows or milk were seen at any one of several cabins visited on a large plantation at some distance from the village where the life was said to be like that of the average plantation negro. The food consisted almost exclusively of fat pork, corn meal, and molasses.

Cooking.—The cooking is of the most simple and primitive character. It is nearly always done over the open fire. Only two of the families visited had stoves. One was that of the carpenter referred to above. He had been under the influence of the Tuskegee Institute. The following extract from a letter of Mr. Hoffman, of the Institute, who shared in the dietary investigations, is of special interest in this connection:

The daily fare is prepared in very simple ways. Corn meal is mixed with water and baked on the flat surface of a hoe or griddle. The salt pork is sliced thin and fried until very brown and much of the grease tried out. Molasses from cane or sorghum is added to the fat, making what is known as "sap," which is eaten with the corn bread. Hot water sweetened with molasses is used as a beverage. This is the bill of fare of most of the cabins on the plantations of the "black belt," three times a day during the year. It is, however, varied at times; thus collards and turnips are boiled with the bacon, the latter being used with the vegetables to supply fat "to make it rich." The corn-meal bread is sometimes made into so-called "cracklin bread," and is prepared as follows: A piece of fat bacon is fried until it is brittle; it is then crushed and mixed with corn meal, water, soda, and salt and baked in an oven over the fireplace. Occasionally the negroes may have an opossum. To prepare this for eating it is first put in hot water to help in removing a part of the hair, then covered with hot ashes until the rest of the hair is removed; thereupon it is put in a large pot, surrounded with sweet potatoes, seasoned with red pepper, and baked. One characteristic of the cooking is that all meats are fried or otherwise cooked until they are crisp. Observation among these people reveals the fact that very many of them suffer from indigestion in some form.

The food and cooking observed in the cabins visited were entirely in accordance with Mr. Hoffman's description, except that flour was used in every case. In how far this was due to the low price which has prevailed of late, and whether the use had extended generally through the black belt of course is not known. It is probable, however, that with the decline in the price of flour the negroes have been learning to use it, and liking its taste and being inclined to imitate the white man in diet as in other things, its use has become more or less common and will be likely to increase.

Clothing.—Of the clothing of the country negroes there is little to be said. It was for the most part coarse, scanty, and ragged. At their work the people did not commonly wear shoes, and for the women a cloth knotted around the head served as a hat whether in the house or the field.

The details of the work at Tuskegee were carried out in the spring of 1895 by Mr. J. W. Hoffman, of the Institute, and Mr. H. M. Smith, special agent of this Department, and in 1896 by Mr. Hoffman. Mr. Green, the farm manager of the Institute, was very helpful in inducing families to allow the investigations to be carried on in their cabins. The whole was under the oversight of Mr. Washington.

The study of individual dietaries generally continued two weeks. On the first day the house was visited, and the pork, meal, flour, molasses, milk, and other food materials on hand were weighed. Each day thereafter a visit was made to the house, and if new materials had been bought meanwhile, they were also weighed. Arrangements were made by which such new materials were kept until they were weighed before any portion was used. As the food was generally purchased only once a week and consisted mainly of fat pork, corn meal, and molasses, the weighing of these articles was a simple matter. With milk, however, especial care had to be taken to insure accurate account of the quantity used. The weighing was done with a large grocers' scale and a small spring balance. At the end of the period of observation an inventory of the food materials on hand was made as at the beginning. The figures of these two inventories, with those of the materials purchased during the study, served for computing the quantities actually consumed. The houses were generally visited once and sometimes twice a day. With a horse and wagon it was not difficult to make the rounds between breakfast time and dark. A considerable number of samples of food materials used were taken for analysis and transmitted to Professor Ross in Auburn, as previously stated.

CALCULATION OF RESULTS.

The quantities of nutrients in the several dietaries were calculated from the weights of the food materials and the proportions of nutrients in each. As it was not found convenient to analyze specimens of all the materials used in each house, enough specimens were selected to give a general idea of the composition, and the composition of the others was assumed from the analyses of these and of other specimens of similar materials. The detailed tables of results show the number of specimens analyzed. As the food consisted almost entirely of salt pork, wheat flour, corn meal, molasses, and milk, materials of tolerably uniform composition, the errors involved in assuming the composition of the specimens not analyzed could hardly be of great importance.

The tabular statement of results beyond give the total quantities of food consumed by each family during the whole period of observation and also the estimated quantities per man per day. These latter estimates are made as follows: It is assumed that a man doing moderately hard muscular work will require on the average a certain amount of nutrients in his daily food, that a woman will eat less, and that young children will eat still less. Counting the amount for the man at

10, the proportions for women and children are taken empirically as follows:

```
Man .................................................................. 10
Woman ..............................................................  8
Boy 14 to 17 years old .............................................  8
Girl 14 to 17 years old ............................................  7
Child 10 to 13 years old ...........................................  6
Child 6 to 9 years old .............................................  5
Child 2 to 5 years old .............................................  4
Child under 2 years old ............................................  3
```

The above ratios accord with such observations of actual food consumption as are available and are used in other estimates of dietaries, the man to be at active and the women and children to have less or no manual labor. In these cases, where the women and children worked in the fields with the men, doubtless larger allowances for their food consumption in comparison with the men would have been more accurate. The ratios are, however, incapable of exact adjustment, and fortunately are of comparatively small importance here.

For various reasons it was found impracticable to collect the waste in the dietary studies. The waste was probably extremely small and for all practical purposes the figures reported may be considered as representing both food purchased and eaten.

DETAILS OF DIETARY STUDIES.

Part of the studies (Nos. 98-105) were made between April 25 and June 20, 1895. The others (Nos. 130-141) were made between December 8, 1895, and February 15, 1896. The details follow:

DIETARY OF A NEGRO CARPENTER'S FAMILY IN ALABAMA (No. 98).

The study began April 25, 1895, and continued fourteen days.

The members of the family and number of meals taken were as follows:

	Meals.
Man about 40 years old	42
Woman about 35 years old (42 meals × 0.8 meal of man) equivalent to	34
Boy 14 years old (42 meals × 0.8 meal of man) equivalent to	34
Boy 12 years old (42 meals × 0.7 meal of man) equivalent to	29
Boy 6 years old (42 meals × 0.5 meal of man) equivalent to	21
Boy 2 years old (42 meals × 0.4 meal of man) equivalent to	17
Total number of meals	177

Equivalent to one man for fifty-nine days.

Remarks.—This family lived in the outskirts of the village of Tuskegee, near the Institute. The father had been under the influence of the latter institution and had learned the carpenter's trade and was in the employ of the Institute. With his savings and labor he had built a very comfortable one-story frame house with four rooms, as shown by the picture (Pl. I, fig. 1). The house was plainly but neatly and very comfortably furnished. A garden supplied the family with vegetables, and two cows and a number of hens and turkeys furnished milk, eggs, and fowl for the table. They had fresh meat frequently, as well as fruits and vegetables. The condition of this family had been steadily improving since the husband came under the influence of the Institute. Instances of such thrift and comfort among the negroes of the region are extremely rare, and were found only in connection with the Institute. They illustrate not what the negro is, but what he may become.

TABLE 4.—*Food materials in dietary No. 98.*

Kind of food material.	Composition.			Weight used.				
	Protein.	Fat.	Carbo-hydrates.	Total cost.	Total food material.	Nutrients.		
						Protein	Fat.	Carbo-hydrates.
ANIMAL FOOD.	*Per ct.*	*Per ct.*	*Per cent.*		*Grams.*	*Grams.*	*Grams.*	*Grams.*
Beef, round [1]	19.4	5.2		$0.10	455	88	24	
Mutton, leg	14.9	14.9		.54	2,040	304	304	
Pork:								
Unsmoked side bacon [1]	8.0	63.2		.51	3,275	262	2,070	
Lard [1]	.3	99.3		.39	2,380	7	2,363	
Total pork and lard				.90	5,655	269	4,433	
Chicken [1]	14.7	6.5		.10	905	133	59	
Eggs [1]	13.2	9.3		.10	595	79	55	
Butter [1]	1.2	82.4		.44	990	12	816 [1]	
Milk [2]	3.5	4.2	5.2	5.04	57,140	2,000	2,400	2,971
Total animal food				7.22	67,780	2,885	8,091	2,971
VEGETABLE FOOD.								
Cereals, sugar, etc.:								
Wheat flour [1]	9.3	.8	78.9	1.40	21,205	1,972	170	16,731
Corn meal [1]	8.5	4.7	73.9	.26	9,300	790	437	6,873
Rolled oats	16.8	7.2	67.0	.01	115	19	8	77
Sugar			100.0	.40	3,020			3.020
Molasses [1]	1.3	.1	68.3	.44	4,410	57	4	3,012
Total cereals and sugar				2.51	38,050	2,838	619	29,713
Fruits:								
Evaporated apples	1.7	2.6	61.3	.08	310	5	8	190
Strawberries	1.0	.7	6.8	.10	665	7	4	45
Total fruits				.18	975	12	12	235
Total vegetable food				2.69	39,025	2,850	631	29,948
Total food				9.91	106,805	5,735	8,722	32,919

[1] Average of analyses of similar Alabama foods.
[2] Fat only determined; protein and carbohydrates calculated as bearing corresponding ratio to average milk.

TABLE 5.—*Weights and percentages of food materials and nutritive ingredients used in dietary No. 98.*

FIG. 1. NEGRO CARPENTER'S HOUSE. (DIETARY NO. 98.)

FIG. 2.—HOUSE AND BARN OF NEGRO FARMER'S FAMILY. (DIETARY NO. 96.)

FIG. 3.—HOUSE, BARN, AND SHEDS OF NEGRO FARMER'S FAMILY. (DIETARIES NOS. 100 AND 130.)

TABLE 5.—*Weights and percentages of food materials and nutritive ingredients used in dietary No. 98—Continued.*

Kind of food material.	Food material.	Nutrients.			Food material.	Nutrients.			Cost.
		Protein.	Fat.	Carbohydrates.		Protein.	Fat.	Carbohydrates.	
FOR FAMILY, 14 DAYS—con.	*Grams.*	*Grams.*	*Grams.*	*Grams.*	*Lbs.*	*Lbs.*	*Lbs.*	*Lbs.*	
Cereals, sugars, starches..	38,050	2,838	619	29,713	83,90	6,30	1.40	65.59	$2.51
Fruits....................	975	12	12	235	2.20	....	.50		.18
Total vegetable food.	39,025	2,850	631	29,948	86.10	6.30	1.40	66.09	2.69
Total food..........	106,805	5,735	8,722	32,919	235.60	12.70	19.20	72.50	9.91
PER MAN PER DAY.									
Beef, veal, and mutton.....	42	7	6	....	.09	.02	.01		
Pork, lard, etc.............	96	5	75	....	.21	.01	.17		
Poultry..................	15	2	1	....	.03	.01			
Eggs...................	10	1	1	....	.02				
Butter.................	17		14	....	.04		.03		
Milk..................	968	34	40	50	2.14	.07	.09	.11	
Total animal food ...	1,148	49	137	50	2.53	.11	.30	.11	.12¼
Cereals, sugars, starches..	645	48	11	504	1.42	.11	.02	1.11	
Fruits.................	17	....		4	.04				
Total vegetable food.	662	48	11	508	1.46	.11	.02	1.11	.04½
Total food..........	1,810	97	148	558	3.99	.22	.32	1.22	.16¾
PERCENTAGES OF TOTAL FOOD.									
	Per ct.	*Per ct.*	*Per ct.*	*Per ct.*					*Per ct.*
Beef, veal, and mutton.....	2.4	6.8	3.8	....					6.5
Pork, lard, etc.............	5.3	4.7	50.8	....					9.1
Poultry..................	.8	2.3	.7	....					1.0
Eggs...................	.6	1.4	.6	....					1.0
Butter.................	.9	.2	9.4	....					4.4
Milk..................	53.5	34.9	27.5	9.0					53.9
Total animal food ...	63.5	50.3	92.8	9.0					72.9
Cereals, sugars, starches..	35.6	49.5	7.1	90.3					25.3
Fruits.................	.9	.2	.1	.7					1.8
Total vegetable food.	36.5	49.7	7.2	91.0					27.1
Total food..........	100.0	100.0	100.0	100.0					100.0

TABLE 6.—*Nutrients and potential energy in food purchased in dietary No. 98.*

Kind of food material.	Cost.	Nutrients.			Fuel value.
		Protein.	Fat.	Carbohydrates.	
FOR FAMILY, 14 DAYS.		*Grams.*	*Grams.*	*Grams.*	*Calories.*
Food purchased:					
Animal.....................	$7.22	2,885	8,091	2,971	99,260
Vegetable..................	2.69	2,850	631	29,948	140,340
Total..................	9.91	5,735	8,722	32,919	239,600
PER MAN PER DAY.					
Food purchased:					
Animal.....................	.12¼	49	137	50	1,680
Vegetable	.04½	48	11	508	2,380
Total..................	.16¾	97	148	558	4,060
PERCENTAGES OF TOTAL FOOD PURCHASED.					
Food purchased:	*Per cent.*	*Per cent.*	*Per cent.*	*Per cent.*	*Per cent*
Animal.....................	72.9	50.3	92.8	9.0	41.4
Vegetable	27.1	49.7	7.2	91.0	58.6
Total..................	100.0	100.0	100.0	100.0	100.0

DIETARY OF A NEGRO FARMER'S FAMILY IN ALABAMA (No. 99).

The study began April 25, 1895, and continued fourteen days.

The members of the family and number of meals taken were as follows:

	Meals.
Man about 55 years old	12
Woman about 50 years old (12 meals × 0.8 meal of man), equivalent to	31
Boy about 17 years old (12 meals × 0.8 meal of man), equivalent to	31
Boy about 16 years old (12 meals × 0.8 meal of man), equivalent to	31
Total number of meals	111

Equivalent to one man for forty-eight days.

Remarks.—This family lived some 2 miles from Tuskegee, and consisted of the husband and wife, about 55 years old, both former slaves, and two boys, aged 17 and 16 years, respectively. One son was learning a trade at the Tuskegee Institute and the other helped on the farm and attended the night school. The man rented about 80 acres of the ordinary "pine land" of the region. The house and farm were remarkably well cared for in comparison with others outside the village (Pl. I, fig. 2). The house, formerly that of a small planter, had five plastered rooms, two of which were rented to a son-in-law. Of their three rooms the family used two as bedrooms and one as a kitchen. In the latter room was a rack about 8 feet high. On this were hung the "sides" of pork which they had slaughtered and salted the previous fall. They had a hogshead of molasses, a barrel of wheat flour, and one of corn meal. All the pork, molasses, and corn used were raised and prepared by the husband. The kitchen was furnished with a small portable cupboard, a pine table, and a few chairs. An iron pot in the fireplace and a frying pan made up the list of cooking utensils. A few plates and other dishes, with knives, forks, and spoons, sufficed for the table, which was without a cloth. The cooking was done in the fireplace, where the bread was baked without yeast or baking powder, and the meat (salt pork) was fried.

In the rear of the house were a barn, a shed, and a small garden.

The live stock consisted of a mule, cows, pigs, and hens. The cows furnished the family with plenty of butter and buttermilk.

The larger part of the farm was devoted to cotton. This, with corn, sugar cane, sorghum, and a few sweet potatoes, made the list of crops. The crops were not mortgaged. The farmer, despite his advanced age, was the most thrifty and progressive man of his class observed in the region.

The wife worked in the field during the busy season.

TABLE 7.—*Food materials in dietary No. 99.*

Kind of food material.	Composition.				Weight used.				
	Pro-tein.	Fat.	Carbo-hydrates.	Total cost.	Total food mate-rial.	Nutrients.			
						Pro-tein.	Fat.	Carbo-hydrates.	
ANIMAL FOOD.	Per ct.	Per ct.	Per cent.		Grams.	Grams.	Grams.	Grams.	
Unsmoked side bacon [1]	8.0	63.2	....	89.74	4.775	382	3,018	....	
Eggs [1]	13.2	9.3	....	.04	245	32	23	....	
Milk [2]	2.6	3.1	3.9	7.07	80,220	2,086	2,480	3.129	
Butter [1]	1.2	82.4	....	.09	200	2	165	....	
Total animal food				7.94	85,440	2,502	5,692	3.129	
VEGETABLE FOOD.									
Cereals, sugar, etc.:									
Wheat flour [1]	9.9	.9	77.7	1.16	17,480	1,731	157	13,582	
Corn meal [1]	7.3	4.1	66.7	.06	2,245	164	92	1,497	
Sugar	...	...	100.0	.05	385	...	...	385	
Molasses [1]	1.3	.1	68.3	.26	2,645	34	3	1,807	
Total vegetable food				1.53	22,755	1,929	252	17.271	
Total food				9.47	108,195	4.431	5,944	20.400	

[1] Average of analyses of similar Alabama foods.
[2] Fat only determined; protein and carbohydrates calculated as bearing corresponding ratio to average milk.

TABLE 8.—*Weights and percentages of food materials and nutritive ingredients used in dietary No. 99.*

Kind of food material.	Food mate-rial.	Nutrients.			Food mate-rial.	Nutrients.			Cost.
		Pro-tein.	Fat.	Carbo-hy-drates.		Pro-tein.	Fat.	Carbo-hy-drates.	
FOR FAMILY, 14 DAYS.	Grams.	Grams.	Grams.	Grams.	Lbs.	Lbs.	Lbs.	Lbs.	
Pork, lard, etc	4.775	382	3,018	....	10.50	0.80	6.70	....	$0.74
Eggs	245	32	23	....	.50	.10	....	....	.04
Butter	200	2	165	....	.50	....	.40	....	.09
Milk	80,220	2,086	2,486	3,129	176.90	4.60	5.50	6.50	7.07
Total animal food	85,440	2,502	5,692	3,129	188.40	5.50	12.60	6.90	7.94
Cereals, sugars, starches	22,755	1,929	252	17,271	50.20	4.30	.50	38.10	1.53
Total vegetable food	22,755	1,929	252	17.271	50.20	4.30	.50	38.10	1.53
Total food	108,195	4,431	5,944	20,400	238.60	9.80	13.10	45.00	9.47
PER MAN PER DAY.									
Pork, lard, etc	99	8	63	....	.22	.02	.14	....	....
Eggs	5	1	1	....	.01	....	....	....	....
Butter	4	...	3	....	.01	....	.01	....	....
Milk	1,672	43	52	65	3.68	.09	.11	.14	....
Total animal food	1,780	52	119	65	3.92	.11	.26	.14	.16½
Cereals, sugars, starches	474	40	5	360	1.05	.09	.01	.80	....
Total vegetable food	474	40	5	360	1.05	.09	.01	.80	.03½
Total food	2,254	92	124	425	4.97	.20	.27	.94	.19½
PERCENTAGES OF TOTAL FOOD.	Per ct.	Per ct.	Per ct.	Per ct.					Per ct.
Pork, lard, etc	4.4	8.7	50.8	....					7.8
Eggs	.2	.7	.4	....					.4
Butter	.2	...	2.8	....					.9
Milk	74.2	47.1	41.8	15.3					74.7
Total animal food	79.0	56.5	95.8	15.3					83.8
Cereals, sugars, starches	21.0	43.5	4.2	84.7					16.2
Total vegetable food	21.0	43.5	4.2	84.7					16.2
Total food	100.0	100.0	100.0	100.0					100.0

TABLE 9.—*Nutrients and potential energy in food purchased in dietary No. 99.*

Kind of food material.	Cost.	Nutrients.			Fuel value.
		Protein.	Fat.	Carbo-hydrates.	
FOR FAMILY, 14 DAYS.					
Food purchased:		*Grams.*	*Grams.*	*Grams.*	*Calories.*
Animal	$7.94	2,502	5,692	3,129	76,020
Vegetable	1.53	1,929	252	17,271	81,060
Total	9.47	4,431	5,944	20,400	157,080
PER MAN PER DAY.					
Food purchased:					
Animal	.16½	52	119	65	1,585
Vegetable	.03¼	40	5	360	1,685
Total	.19¾	92	124	425	3,270
PERCENTAGES OF TOTAL FOOD PURCHASED.					
Food purchased:		*Per cent.*	*Per cent.*	*Per cent.*	*Per cent.*	*Per cent.*
Animal	83.8	56.5	95.8	15.3	48.4
Vegetable	16.2	43.5	4.2	84.7	51.6
Total	100.0	100.0	100.0	100.0	100.00

DIETARY OF A NEGRO FARMER'S FAMILY IN ALABAMA (No. 100).

The study began April 25, 1895, and continued fourteen days.
The members of the family and number of meals taken were as follows:

	Meals.
Man about 35 years old	42
Woman about 35 years old (42 meals × 0.8 meal of man) equivalent to	34
Child 11 years old (42 meals × 0.6 meal of man) equivalent to	25
Child 7 years old (42 meals × 0.5 meal of man) equivalent to	21
Child 4 years old (42 meals × 0.4 meal of man) equivalent to	17
Two children under 2 years of age (84 meals × 0.3 meal of man) equivalent to	25
Man visitor	14
Total number of meals	178

Equivalent to one man for fifty-nine days.

Remarks.—This family was composed of husband and wife and five children. The cabin was built of logs and had two rooms (Pl. I, fig. 3). One, used as living and sleeping room, contained two beds and a few small pieces of furniture. The kitchen was provided with a pine table, one or two chairs, a small portable cupboard, the usual pot and frying pan, and a few dishes for the table. There was no churn, as the family had no cow. In the cupboard were a piece of salt pork and a jug of molasses, and near by a sack of corn meal. The provisions were purchased each week, and toward the close there was very little left in the house. Fried pork and corn pone, cooked in the fireplace, composed the daily diet.

A mule, an ox, and a pig made up the live stock.

The farm was planted chiefly to cotton. A small patch was devoted to sugar cane. There was no garden, and the cotton was cultivated close up to the cabin door.

This farmer had been in the habit of mortgaging his crops each year.

but under the influence of the Institute and the farmers' conferences he was trying to better his condition and was working this year without a mortgage.

TABLE 10.—*Food materials in dietary No. 100.*

Kind of food material.	Composition.			Total cost.	Total food material.	Weight used.		
	Protein.	Fat.	Carbohydrates.			Nutrients.		
						Protein.	Fat.	Carbohydrates.
ANIMAL FOOD.	Per ct.	Per ct.	Per cent.		Grams.	Grams.	Grams.	Grams.
Unsmoked side bacon¹	8.9	63.2	....	$0.25	1,590	127	1,005	....
Lard¹	.3	99.3	....	.24	1,445	4	1,435	....
Total animal food				.49	3,035	131	2,440	....
VEGETABLE FOOD.								
Cereals, sugar, etc.:								
Wheat flour¹	9.6	.8	78.3	.63	9,470	909	76	7,415
Corn meal¹	7.3	4.1	66.7	.58	20,920	1,527	858	13,954
Rice¹	7.5	.4	79.3	.08	710	53	3	563
Total cereals				1.29	31,100	2,489	937	21,932
Collards (cabbage)¹	2.2	.4	5.7	.01	255	6	1	15
Total vegetable food				1.30	31,355	2,495	938	21,947
Total food				1.79	34,390	2,626	3,378	21,947

¹Average of analyses of similar Alabama foods.

TABLE 11.—*Weights and percentages of food materials and nutritive ingredients used in dietary No. 100.*

Kind of food material.	Food material.	Nutrients.			Food material.	Nutrients.			Cost.
		Protein.	Fat.	Carbohydrates.		Protein.	Fat.	Carbohydrates.	
FOR FAMILY, 14 DAYS.	Grams.	Grams.	Grams.	Grams.	Lbs.	Lbs.	Lbs.	Lbs.	
Pork, lard, etc	3,035	131	2,440	....	6.70	0.30	5.40	....	$0.49
Cereals, sugars, starches	31,100	2,489	937	21,932	68.50	5.50	2.10	48.40	1.29
Vegetables	255	6	1	15	.60	....	....	....	.01
Total vegetable food	31,355	2,495	938	21,947	69.10	5.50	2.10	48.40	1.30
Total food	34,390	2,626	3,378	21,947	75.80	5.80	7.50	48.40	1.79
PER MAN PER DAY.									
Pork, lard, etc	51	2	41	....	.11	.01	.09	....	
Total animal food	51	2	41	....	.11	.01	.09	....	.003
Cereals, sugars, starches	527	42	16	372	1.16	.09	.04	.82	
Vegetables	5	4	....	....	.01				
Total vegetable food	532	42	16	372	1.17	.09	.04	.82	.023
Total food	583	44	57	372	1.28	.10	.13	.82	.03
PERCENTAGES OF TOTAL FOOD.	Per ct.	Per ct.	Per ct.	Per ct.					Per ct.
Pork, lard, etc	8.8	5.0	72.2	....					27.4
Total animal food	8.8	5.0	72.2	....					27.4
Cereals, sugars, starches	90.4	94.8	27.8	99.9					72.1
Vegetables	.8	.2	....	.1					.5
Total vegetable food	91.2	95.0	27.8	100.0					72.6
Total food	100.0	100.0	100.0	100.0					100.0

TABLE 12.—*Nutrients and potential energy in food purchased in dietary No. 100.*

| Kind of food material. | Cost. | Nutrients. | | | Fuel value. |
| | | Protein. | Fat. | Carbo-hydrates. | |

FOR FAMILY, 14 DAYS.

		Grams.	*Grams.*	*Grams.*	*Calories.*
Food purchased:					
Animal	$0.49	131	2,440	....	23,230
Vegetable	1.30	2,495	938	21,947	108,930
Total	1.79	2,626	3,378	21,947	132,160

PER MAN PER DAY.

Food purchased:					
Animal	.00⅜	2	41	....	395
Vegetable	.02¼	42	16	372	1,845
Total	.03	44	57	372	2,240

PERCENTAGES OF TOTAL FOOD PURCHASED.

	Per cent.	*Per cent.*	*Per cent.*	*Per cent.*	*Per cent.*
Food purchased:					
Animal	27.4	5.0	72.2	....	17.6
Vegetable	72.6	95.0	27.8	100.0	82.4
Total	100.0	100.0	100.0	100.0	100.0

DIETARY OF A NEGRO FARMER'S FAMILY IN ALABAMA (No. 101).

The study began April 26, 1895, and continued seven days.
The members of the family and number of meals taken were as follows:

Meals.
Man about 60 years old 21
Woman about 55 years old (21 meals × 0.8 meal of man) equiva-
lent to .... 17 ·
Two children between 10 and 14 years old (42 meals × 0.6 meal of
man) 25
Child between 6 and 10 years old (21 meals × 0.5 meal of man)
equivalent to 11
Child 4 years old (21 meals × 0.4 meal of man) equivalent to 8
Man visitor 15

Total number of meals 97
Equivalent to one man for thirty-two days.

Remarks.—This family consisted of six persons—the husband, wife,
and four children. A mortgage had caused them to be sold out the
year previous, and they were now clearing up a new place on a piece of
land which was more than a mile back from the traveled road, and of
which only a part had previously been cultivated. The cabin, however,
was better than the majority. It contained four rooms, and a small log
hut in the yard served as a kitchen (Pl. II, fig. 1. The soil was sandy
and very poor. The land was partly covered with pine trees, full of
stumps and second growth. Cotton, cane, sweet potatoes, and cowpeas
had been planted. The woman worked all day in the field.

The live stock consisted of a mule, two cows, and some hens. Milk
and eggs were used, and occasionally the family indulged in the luxury
of sugar and coffee. The provisions were purchased by the week. The
cooking was done with a stove, and the diet was somewhat better than
that in several other cabins where studies were made.

FIG. 1. HOUSE OF NEGRO FARMER'S FAMILY. (DIETARY NO. 101.)

FIG. 2. HOUSE OF NEGRO SAWMILL LABORER. (DIETARIES NOS. 102 AND 131.)

FIG. 3. HOUSE OF NEGRO COTTON PLANTATION LABORER. (DIETARY NO. 103.)

All the possessions of the family were under mortgage, but their three oldest sons were working for the mortgagee to assist in payment. This accounts for their having such a good cabin after the previous year's reverses.

TABLE 13.—*Food materials in dietary No. 101.*

Kind of food material.	Composition.				Weight used.			
	Protein.	Fat.	Carbohydrates.	Total cost.	Total food material.	Nutrients.		
						Protein.	Fat.	Carbohydrates.
ANIMAL FOOD.								
Pork:	*Per ct.*	*Per ct.*	*Per cent.*		*Grams.*	*Grams.*	*Grams.*	*Grams.*
Bacon [1]	8.0	63.2	....	$0.52	2,990	239	1,890	....
Lard [1]	.3	99.3	....	.15	905	3	899	....
Total pork and lard	....	....	....	.67	3,895	242	2,789	....
Eggs [1]	13.2	9.3	....	.13	640	84	60	....
Butter [1]	1.2	82.4	....	.26	595	7	490	....
Milk [2]	3.8	4.6	5.7	.39	4,125	168	204	252
Buttermilk	3.0	.5	4.8	.34	12,460	374	62	598
Total animal food	....	....	....	1.79	22,015	875	3,605	850
VEGETABLE FOOD.								
Cereals, sugar, etc.:								
Wheat flour [1]	9.6	.8	78.3	.56	8,390	805	67	6,569
Corn meal [1]	7.3	4.1	66.7	.23	8,280	604	339	5,523
Sugar (C)	....	....	95.0	.10	905	....	....	860
Total vegetable food	....	....	....	.89	17,575	1,409	406	12,952
Total food	....	....	....	2.68	39,590	2,284	4,011	13,802

[1] Average of analyses of similar Alabama foods.
[2] Fat only determined; protein and carbohydrates calculated as bearing corresponding ratio to average milk.

TABLE 14.—*Weights and percentages of food materials and nutritive ingredients used in dietary No. 101.*

Kind of food material.	Food material.	Nutrients.			Food material.	Nutrients.			Cost.
		Protein.	Fat.	Carbohydrates.		Protein.	Fat.	Carbohydrates.	
FOR FAMILY, 7 DAYS.									
	Grams.	*Grams.*	*Grams.*	*Grams*	*Lbs.*	*Lbs.*	*Lbs.*	*Lbs.*	
Pork, lard, etc	3,895	242	2,789	....	8.60	0.50	6.20	....	$0.67
Eggs	640	84	60	....	1.4	.20	.10	....	.13
Butter	595	7	490	....	1.30	....	1.10	....	.26
Milk	4,425	168	204	252	9.70	.40	.50	0.60	.39
Buttermilk	12,460	374	62	598	27.50	.80	....	1.30	.34
Total animal food	22,015	875	3,605	850	48.50	1.90	7.90	1.90	1.79
Cereals, sugars, starches	17,575	1,409	406	12,952	38.80	3.10	.90	28.50	.89
Total vegetable food	17,575	1,409	406	12,952	38.80	3.10	.90	28.50	.89
Total food	39,590	2,284	4,011	13,802	87.30	5.00	8.80	30.40	2.68
PER MAN PER DAY.									
Pork, lard, etc	122	7	87	....	.27	.02	.19	....	
Eggs	20	3	2	....	.04	.01	....	....	
Butter	19	....	15	....	.04	....	.03	....	
Milk	138	5	7	8	.30	.01	.02	.02	
Buttermilk	380	12	2	19	.86	.02	....	.04	
Total animal food	688	27	113	27	1.51	.06	.24	.06	.05¼
Cereals, sugars, starches	549	44	13	405	1.21	.10	.03	.89	
Total vegetable food	549	44	13	405	1.21	.10	.03	.89	.02¾
Total food	1,237	71	126	432	2.72	.16	.27	.95	.08¼

32

TABLE 14.—*Weights and percentages of food materials and nutritive ingredients used in dietary No. 101—Continued.*

Kind of food material.	Food material.	Nutrients.			Food material.	Nutrients.			Cost.
		Protein.	Fat.	Carbohydrates.		Protein.	Fat.	Carbohydrates.	
PERCENTAGES OF TOTAL FOOD.	Per ct.	Per ct.	Per ct.						Per ct.
Pork, lard, etc	9.8	10.6	69.6						25.0
Eggs	1.6	3.7	1.5						4.8
Butter	1.5	.3	12.2						9.7
Milk	11.2	7.3	5.1	1.9					14.6
Buttermilk	31.5	16.4	1.5	4.3					12.7
Total animal food	55.6	38.3	89.9	6.2					66.8
Cereals, sugars, starches	44.4	61.7	10.1	93.8					33.2
Total vegetable food	44.4	61.7	10.1	93.8					33.2
Total food	100.0	100.0	100.0	100.0					100.0

TABLE 15.—*Nutrients and potential energy in food purchased in dietary No. 101.*

Kind of food material.	Cost.	Nutrients.			Fuel value.
		Protein.	Fat.	Carbohydrates.	
FOR FAMILY, 7 DAYS.					
Food purchased:		Grams.	Grams.	Grams.	Calories.
Animal	$1.79	875	3,605	850	40,600
Vegetable	.89	1,409	406	12,952	62,655
Total	2.68	2,284	4,011	13,802	103,255
PER MAN PER DAY.					
Food purchased:					
Animal	.05¼	27	113	27	1,270
Vegetable	.02¾	44	13	405	1,960
Total	.08¼	71	126	432	3,230
PERCENTAGE OF TOTAL FOOD PURCHASED.					
Food purchased:	Per cent.	Per cent.	Per cent.	Per cent.	Per cent.
Animal	66.8	38.3	89.9	6.2	39.3
Vegetable	33.2	61.7	10.1	93.8	60.7
Total	100.0	100.0	100.0	100.0	100.0

DIETARY OF A NEGRO SAWMILL LABORER'S FAMILY IN ALABAMA (No. 102).

The study began April 27, 1895, and continued seven days.
The members of the family and number of meals taken were as follows:

	Meals.
Man about 26 years old	21
Woman about 24 years old (21 meals × 0.8 meal of man) equivalent to	17
Girl 12 years old (21 meals × 0.6 meal of man) equivalent to	13
Two children between 2 and 4 years old (12 meals × 0.4 meal of man) equivalent to	17
Total number of meals	68

Equivalent to one man for twenty-three days.

Remarks.—This family, living on a small farm, consisted of husband and wife, an adopted daughter of 12, and two younger children. The husband worked in a sawmill and received 50 cents a day, paid in merchandise. The wife worked in the field. The husband was a step-son of the family mentioned in study No. 101. The two farms joined, and the people assisted one another in their farm work.

The cabin was made of logs and had one room, which served as living and sleeping room, and an addition in the rear, not more than 6 feet wide, which was used as a kitchen (Pl. II, fig. 2). The living room contained two beds, a few other small pieces of furniture, and a fireplace. The kitchen had a cupboard, a pine table, and a fireplace. The house had two doors, but was without windows. The cabin was comparatively new, and better than many of its class. The furnishings, however, were as poor as those of the ordinary one-room cabins.

The food consisted of salt pork, corn meal, and molasses, with butter, buttermilk, "clabber," and a few eggs. For live stock the people had an ox, a cow, and a few hens. They had no garden, the farm crops were like those of No. 101, cotton, with a little cane, cowpeas, and sweet potatoes.

TABLE 16.—*Food materials in dietary No. 102.*

Kind of food material.	Composition.					Weight used.		
	Protein.	Fat.	Carbohydrates.	Total cost.	Total food material.	Nutrients.		
						Protein.	Fat.	Carbohydrates.
ANIMAL FOOD.	Per ct.	Per ct.	Per cent.		Grams.	Grams.	Grams.	Grams.
Bacon [1]	8.0	63.2	....	$0.29	1,630	130	1,030	....
Eggs [1]	13.2	9.3	....	.07	425	56	40	....
Butter [1]	1.2	82.4	....	.11	240	3	198	....
Buttermilk [1]	3.0	.5	4.8	.92	1,815	54	9	87
Milk [1]	3.3	4.0	5.0	.05	10,435	344	417	5.2
Total animal food				1.44	14,545	5-7	1,694	609
VEGETABLE FOOD.								
Cereals, sugar, etc.:								
Wheat flour [1]	9.6	.8	78.3	.24	3,645	350	29	2,854
Corn meal [1]	7.3	4.1	66.7	.14	5,100	372	209	3,402
Sugar (C)	....	....	95.0	.06	540	....	....	513
Molasses [1]	1.3	.1	68.3	.36	3,670	47	4	2,507
Total vegetable food				.80	12,955	769	242	9,276
Total food				2.24	27,500	1,356	1,936	9,885

[1] Average of analyses of similar Alabama foods.

34

TABLE 17.—*Weights and percentages of food materials and nutritive ingredients used in dietary No. 102.*

Kind of food material.	Food material.	Nutrients.			Food material.	Nutrients.			Cost.
		Protein.	Fat.	Carbohydrates.		Protein.	Fat.	Carbohydrates.	

FOR FAMILY, 7 DAYS.

	Grams.	Grams.	Grams.	Grams.	Lbs.	Lbs.	Lbs.	Lbs.	
Pork, lard, etc	1,630	130	1,630	....	3.60	0.30	2.30	....	$0.29
Eggs	425	56	40	....	1.00	.10	.10	....	.07
Butter	240	3	198	....	.50	....	.50	....	.11
Milk	10,435	344	417	522	23.00	.80	.90	1.10	.92
Buttermilk	1,845	54	9	87	4.00	.10	....	.20	.05
Total animal food	14,545	587	1,694	609	32.10	1.30	3.80	1.30	1.44
Cereals, sugars, starches	12,955	769	242	9,276	28.50	1.70	.50	20.50	.80
Total vegetable food	12,955	769	242	9,276	28.50	1.70	.50	20.50	.80
Total food	27,500	1,356	1,936	9,885	60.60	3.00	4.30	21.80	2.24

PER MAN PER DAY.

Pork, lard, etc	71	6	45	....	.16	.01	.10	....	....
Eggs	18	3	2	....	.04	.01	....	....	....
Butter	10	....	9	....	.02	....	.02	....	....
Milk	454	15	18	23	1.00	.03	.04	.05	....
Buttermilk	79	2	....	3	.17	.01	....	.01	.06¼
Total animal food	632	26	74	26	1.39	.06	.16	.06	....
Cereals, sugars, starches	563	33	11	403	1.25	.07	.02	.89	....
Total vegetable food	563	33	11	403	1.25	.07	.02	.89	.01½
Total food	1,195	59	85	429	2.64	.13	.18	.95	.09¾

PERCENTAGES OF TOTAL FOOD.

	Per ct.	Per ct.	Per ct.	Per ct.					Per ct.
Pork, lard, etc	5.9	9.6	53.2	....					13.0
Eggs	1.5	4.1	2.1	....					3.1
Butter	.9	.2	10.3	....					4.9
Milk	38.0	25.4	21.5	5.3					41.1
Buttermilk	6.6	4.0	.4	.9					2.2
Total animal food	52.9	43.3	87.5	6.2					64.3
Cereals, sugars, starches	47.1	56.7	12.5	93.8					35.7
Total vegetable food	47.1	56.7	12.5	93.8					35.7
Total food	100.0	100.0	100.0	100.0					100.0

TABLE 18.—*Nutrients and potential energy in food purchased in dietary No. 102.*

Kind of food material.	Cost.	Nutrients.			Fuel value.
		Protein.	Fat.	Carbohydrates.	

FOR FAMILY, 7 DAYS.

Food purchased:		Grams.	Grams.	Grams.	Calories.
Animal	$1.44	587	1,694	609	20,660
Vegetable	.80	769	242	9,276	43,435
Total	2.24	1,356	1,936	9,885	64,095

PER MAN PER DAY.

Food purchased:					
Animal	.06¼	26	74	26	900
Vegetable	.03½	33	11	403	1,890
Total	.09¾	59	85	429	2,790

PERCENTAGES OF TOTAL FOOD PURCHASED.

Food purchased:	Per cent.	Per cent.	Per cent.	Per cent.	Per cent.
Animal	64.3	43.3	87.5	6.2	32.2
Vegetable	35.7	56.7	12.5	93.8	67.8
Total	100.0	100.0	100.0	100.0	100.0

DIETARY OF A NEGRO COTTON PLANTATION LABORER'S FAMILY IN ALABAMA
(No. 103).

The study began May 12, 1895, and continued fourteen days.
The members of the family and number of meals taken were as follows:

	Meals.
Man about 30 years old	42
Woman about 80 years old (12 meals × 0.8 meal of man) equivalent to	31
Woman about 40 years old (42 meals × 0.8 meal of man) equivalent to	34
Child 16 years old (12 meals × 0.8 meal of man) equivalent to	34
Child 14 years old (42 meals × 0.7 meal of man) equivalent to	29
Child 3 years old (42 meals × 0.4 meal of man) equivalent to	17
Child 1 year old (12 meals × 0.3 meal of man) equivalent to	12
Total number of meals	202

Equivalent to one man for sixty-seven days.

Remarks.—This family consisted of the husband, wife, another woman quite old, and four children.

They occupied a log cabin of two rooms on a plantation of 1,200 acres, 8 miles from the village (Pl. II, fig. 3). The plantation was divided into small farms averaging 40 acres, and rented to some thirty families. This family bought their supplies from the proprietor, and gave him as security a mortgage on their crops. The provisions, instead of being purchased by the week, as was so commonly done, were bought in larger quantities once in four or six weeks, and stored in a log hut close by the house. The provender for the mules was likewise bought in quantity. With the exception of a small vegetable garden of collards, the farm was planted entirely with cotton.

The live stock consisted of two mules and three hogs.

TABLE 19.—*Food materials in dietary No. 103.*

Kind of food material.	Composition.			Total cost.	Weight used.			
	Protein.	Fat.	Carbohydrates.		Total food material.	Nutrients.		
						Protein.	Fat.	Carbohydrates.
ANIMAL FOOD.	Per ct.	Per ct.	Per cent.		Grams.	Grams.	Grams.	Grams.
Bacon [1]	8.0	63.2		$4.87	27,615	2,209	17,453	
Total animal food				4.87	27.615	2,209	17,453	
VEGETABLE FOOD.								
Cereals, sugar, etc.:								
Corn meal [1]	7.3	4.1	66.7	.97	32,580	2,379	1,336	21,738
Wheat flour [1]	9.6	.8	78.3	.90	14,740	1,415	118	11,511
Molasses [1]	1.3	.1	68.3	1.49	14,970	195	15	10,224
Total vegetable food				3.36	62,300	3,989	1,469	43,503
Total food				8.23	89,915	6,198	18,922	43,503

[1] Average of analyses of similar Alabama foods.

TABLE 20.—*Weights and percentages of food materials and nutritive ingredients used in dietary No. 103.*

Kind of food material.	Food material.	Nutrients.			Food material.	Nutrients.			Cost.
		Protein.	Fat.	Carbohydrates.		Protein.	Fat.	Carbohydrates.	
FOR FAMILY, 14 DAYS.	Grams.	Grams.	Grams.	Grams.	Lbs.	Lbs.	Lbs.	Lbs.	
Pork, lard, etc	27,615	2,209	17,453	........:.	60.90	4.90	38.50	....	$4.87
Cereals, sugars, starches..	62,300	3,989	1,469	43,503	137.30	8.80	3.20	95.90	3.36
Total food	89,915	6,198	18,922	43,503	198.20	13.70	41.70	95.90	8.23
PER MAN PER DAY.									
Pork, lard, etc	412	33	261	....	.91	.07	.57	....	.07¼
Cereals, sugars, starches..	930	60	22	649	2.05	.13	.05	1.43	.05
Total food	1,342	93	283	649	2.96	.20	.62	1.43	.12¼
PERCENTAGES OF TOTAL FOOD.	Per ct.	Per ct.	Per ct.	Per ct.					Per ct.
Pork, lard, etc	30.7	35.6	92.3	....					59.2
Cereals, sugars, starches..	69.3	64.4	7.7	100.0					40.8
Total food	100.0	100.0	100.0	100.0					100.0

TABLE 21.—*Nutrients and potential energy in food purchased in dietary No. 103.*

Kind of food material.	Cost.	Nutrients.			Fuel value.
		Protein.	Fat.	Carbohydrates.	
FOR FAMILY, 14 DAYS.					
Food purchased:		Grams.	Grams.	Grams.	Calories.
Animal	$4.87	2,209	17,453	....	171,370
Vegetable	3.36	3,989	1,469	43,503	208,380
Total	8.23	6,198	18,922	43,503	379,750
PER MAN PER DAY.					
Food purchased:					
Animal	.07¼	33	261	....	2,560
Vegetable	.05	60	22	649	3,110
Total	.12¼	93	283	649	5,670
PERCENTAGES OF TOTAL FOOD PURCHASED.					
Food purchased:	Per cent.	Per cent.	Per cent.	Per cent.	Per cent.
Animal	59.2	35.6	92.3	....	45.1
Vegetable	40.8	64.4	7.7	100.0	54.9
Total	100.0	100.0	100.0	100.0	100.0

DIETARY OF A NEGRO COTTON PLANTATION LABORER'S FAMILY IN ALABAMA (No. 104).

The study began May 12, 1895 and continued fourteen days.

The members of the family and number of meals taken were as follows:

	Meals.
Man about 45 years old	12
Man 22 years old	12
Woman about 50 years old (42 meals × 0.8 meal of man) equivalent to	34
Girl 16 years old (42 meals × 0.7 meal of man) equivalent to	29
Total number of meals	117

Equivalent to one man for forty-nine days.

Remarks.—This family was composed of husband, wife, and two grown children. They lived on the same plantation as the family in dietary No. 103 and their condition was very similar. They, however, purchased their provisions in small quantities by the week. The cabin was situated on the bank of a creek bordering a swamp, and a cupboard, meal barrel, a few chairs, a pine table, and two beds furnished the house. Their live stock consisted of one mule and a pig. The farm was planted entirely to cotton. All the women worked in the field.

TABLE 22.—*Food materials in dietary No. 104.*

Kind of food material.	Composition.					Weight used.		
	Pro- tein.	Fat.	Carbo- hydrates.	Total cost.	Total food mate- rial.	Nutrients.		
						Pro- tein.	Fat.	Carbo- hydrates.
ANIMAL FOOD.	*Per ct.*	*Per ct.*	*Per cent.*		*rams.*	*Grams.*	*Grams.*	*Grams.*
Bacon [1]	8.0	63.2	....	$3.16	17,915	1,433	11,322	....
Total animal food				3.16	17,915	1,433	11,322	....
VEGETABLE FOOD.								
Cereals, sugar, etc.:								
Corn meal [1]	7.3	4.1	66.7	1.26	21,005	1,533	861	14,010
Wheat flour [1]	9.6	.8	78.3	.58	19,050	1,829	152	14,916
Molasses [1]	1.3	.1	68.3	.54	5,430	71	5	3,709
Total vegetable food				2.38	45,485	3,433	1,018	32,635
Total food				5.54	63,400	4,866	12,340	32,635

[1] Average of analyses of similar Alabama foods.

TABLE 23.—*Weights and percentages of food materials and nutritive ingredients used in dietary No. 104.*

Kind of food material.	Food mate- rial.	Nutrients.			Food mate- rial.	Nutrients.			Cost.
		Pro- tein.	Fat.	Carbo- hy- drates.		Pro- tein.	Fat.	Carbo- hy- drates.	
FOR FAMILY, 14 DAYS.	*Grams.*	*Grams.*	*Grams.*	*Grams.*	*Lbs.*	*Lbs.*	*Lbs.*	*Lbs.*	
Pork, lard, etc	17,915	1,433	11,322	....	39.50	3.10	25,00	....	$3.16
Cereals, sugars, starches	45,485	3,433	1,018	32,635	100.30	7.60	2.20	72.00	2.38
Total food	63,400	4,866	12,340	32,635	139.80	10.70	27.20	72.00	5.54
PER MAN PER DAY.									
Pork, lard, etc	366	29	231	....	.81	.06	.51	....	.06¼
Cereals, sugars, starches	928	70	21	666	2.04	.15	.05	1.47	.04¾
Total food	1,294	99	252	666	2.85	.21	.56	1.47	.11¼
PERCENTAGES OF TOTAL FOOD.									
	Per ct.	*Per ct.*	*Per ct.*	*Per ct.*					*Per ct.*
Pork, lard, etc	28.3	29.4	91.8	....					57.0
Cereals, sugars, starches	71.7	70.6	8.2	100.0					43.0
Total food	100.0	100.0	100.0	100.0					100.0

TABLE 24.—*Nutrients and potential energy in food purchased in dietary No. 104.*

Kind of food material.	Cost.	Nutrients.			Fuel value.
		Protein.	Fat.	Carbo-hydrates.	
FOR FAMILY, 14 DAYS.		*Grams.*	*Grams.*	*Grams.*	*Calories.*
Food purchased:					
Animal	$3.16	1,433	11,322	....	111,170
Vegetable	2.38	3,433	1,018	32,635	157,345
Total	5.54	4,866	12,340	32,635	268,515
PER MAN PER DAY.					
Food purchased:					
Animal	.06⅜	29	231	....	2,270
Vegetable	.04⅞	70	21	666	3,210
Total	.11⅜	99	252	666	5,480
PERCENTAGES OF TOTAL FOOD PURCHASED.					
Food purchased:		*Per cent.*	*Per cent.*	*Per cent.*	*Per cent.*	*Per cent.*
Animal	57.0	29.5	91.8	....	41.4
Vegetable	43.0	70.5	8.2	100.0	58.6
Total	100.0	100.0	100.0	100.0	100.0

DIETARY OF A NEGRO FARM MANAGER'S FAMILY IN ALABAMA (No. 105).

The study began June 10, 1895, and continued ten days.
The members of the family and number of meals taken were as follows:

	Meals.
Man about 41 years old	30
Woman 22 years old (30 meals × 0.8 meal of man) equivalent to	24
Woman 22 years old (30 meals × 0.8 meal of man) equivalent to	24
Woman 23 years old (30 meals × 0.8 meal of man) equivalent to	24
Boy 13 years old (30 meals × 0.7 meal of man) equivalent to	21
Total number of meals	123

Equivalent to one man for forty-one days.

Remarks.—This family consisted of a husband, wife, two women, one a niece of the family and the other a visitor, and a boy. They lived on the Tuskegee Institute farm, of which the husband was manager, in a five-room cottage. Their home was furnished in modern style, simply yet neatly. The wife was one of the teachers in the Institute. The niece, 22 years old, assisted in housework; the boy attended school and did chores around the house; the visitor was a dressmaker. The cooking was plain. Plenty of oil and fat was used in the preparation of their food, as is the case in nearly all Southern cooking.

TABLE. 25.—*Food materials in dietary No. 105.*

Kind of food material.	Composition.				Weight used.			
	Protein.	Fat.	Carbohydrates.	Total cost.	Total food material.	Nutrients.		
						Protein.	Fat.	Carbohydrates.
ANIMAL FOOD.	*Per ct.*	*Per ct.*	*Per cent.*		*Grams.*	*Grams.*	*Grams.*	*Grams.*
Beef, round steak [1]	19.4	5.2	....	$0.18	795	154	41	....
Pork:								
Bacon [1]	8.0	63.2	....	.59	3,345	268	2,114	....
Lard [1]	.3	99.3	....	.47	2,835	9	2,815	....
Total	....	....	....	1.06	6,180	277	4,929	....
Poultry, chicken [1]	14.7	6.5	....	.12	1,105	162	72	....
Eggs [1]	13.2	9.3	....	.11	595	79	55	....
Butter [1]	1.2	82.4	....	.20	455	5	375	....
Milk	3.3	4.0	5.0	.04	480	16	19	24
Buttermilk [1]	3.0	.5	4.8	.13	4,875	146	24	234
Total animal food	....	....	....	1.84	14,485	839	5,515	2'8
VEGETABLE FOOD.								
Cereals, sugar, etc.:								
Flour, wheat [1]	9.6	.8	78.3	.60	9,070	871	73	7,102
Molasses [1]	1.3	.1	68.3	.27	2,720	35	3	1,858
Total	....	....	....	.87	11,790	906	76	8,960
Vegetables:								
Beans, string	2.2	.4	9.4	.03	1,190	26	5	112
Beets	1.6	.1	9.6	.22	1,955	31	2	184
Cabbage	2.1	.4	5.8	.70	6,350	133	25	368
Corn, green	2.8	1.1	14.1	.05	1,020	29	11	114
Okra	2.0	.4	9.5	....	115	2	5	11
Onions	1.7	.4	9.9	.05	945	16	4	93
Tomatoes	.8	.4	3.9	.02	795	6	3	31
Total	....	....	....	1.07	12,360	243	55	947
Fruits, nuts, etc.:								
Blackberries	.9	2.1	7.5	.66	1,360	12	29	192
Peaches	1.1	....	13.4	.02	1,305	14	....	175
Total	....	....	....	.68	2,665	26	29	277
Total vegetable food	....	....	....	2.02	26,815	1,175	160	10,184
Total food	....	....	....	3.86	41,300	2,014	5,675	10,442

[1] Average of analyses of similar Alabama foods.

TABLE 26.—*Weights and percentages of food materials and nutritive ingredients used in dietary No. 105.*

Kind of food material.	Food material.	Nutrients.			Food material.	Nutrients.			Cost.
		Protein.	Fat.	Carbohydrates.		Protein.	Fat.	Carbohydrates.	
FOR FAMILY, 10 DAYS.	*Grams.*	*Grams.*	*Grams.*	*Grams.*	*Lbs.*	*Lbs.*	*Lbs.*	*Lbs.*	
Beef, veal, and mutton	795	154	41	....	1.80	0.30	0.10	....	$0.18
Pork, lard, etc	6,180	277	4,929	....	13.60	.60	10.89	....	1.06
Poultry	1,105	162	72	....	2.40	.40	.20	....	.12
Eggs	595	79	55	....	1.30	.20	.10	....	.11
Butter	455	5	375	....	1.00	....	.80	....	.20
Milk	480	16	19	24	1.10	.10	.10	0.10	.04
Buttermilk	4,875	146	24	234	10.70	.30	....	.50	.13
Total animal food	14,485	839	5,515	258	31.90	1.90	12.10	.60	1.84
Cereals, sugars, starches	11,790	906	76	8,960	26.00	2.00	.20	19.80	.87
Vegetables	12,360	243	55	947	27.20	.50	.10	2.10	1.07
Fruits	2,665	26	29	277	5.90	.10	.10	.60	.04
Total vegetable food	26,815	1,175	160	10,184	59.10	2.60	.40	22.50	2.02
Total food	41,300	2,014	5,675	10,442	91.00	4.50	12.50	23.10	3.86

TABLE 26.—*Weights and percentages of food materials and nutritive ingredients used in dietary No. 105—Continued.*

Kind of food material.	Food material.	Nutrients.			Food material.	Nutrients.			Cost.
		Protein.	Fat.	Carbohydrates.		Protein.	Fat.	Carbohydrates.	

PER MAN PER DAY.

	Grams.	Grams.	Grams.	Grams.	Lbs.	Lbs.	Lbs.	Lbs.	
Beef, veal, and mutton	19	4	1		0.04	0.01			
Pork, lard, etc	151	7	120		.34	.02	0.27		
Poultry	27	4	2		.06	.01			
Eggs	14	2	1		.03				
Butter	11		9		.02		.02		
Milk	12				.03				
Buttermilk	119	3	1	6	.26	.01		0.01	
Total animal food	353	20	134	6	.78	.05	.29	.01	$0.04½
Cereals, sugars, starches	288	22	2	219	.63	05	.01	.48	
Vegetables	301	6	1	23	.67	.01		.06	
Fruits	65	1	1	7	.14			.01	
Total vegetable food	654	29	4	249	1.44	.06	.01	.55	.05
Total food	1,007	49	138	255	2.22	.11	.30	.56	.09½

PERCENTAGES OF TOTAL FOOD.

	Per ct.	Per ct.	Per ct.	Per ct.					Per ct.
Beef, veal, and mutton	1.9	7.7	0.7						4.7
Pork, lard, etc	15.0	13.8	86.9						27.5
Poultry	2.7	8.0	1.3						3.1
Eggs	1.4	3.9	1.0						2.8
Butter	1.1	.2	6.6						5.2
Milk	1.2	.8	.3	.2					1.0
Buttermilk	11.8	7.3	.4	2.2					3.4
Total animal food	35.1	41.7	97.2	2.4					47.7
Cereals, sugars, starches	28.6	45.0	1.3	85.8					22.5
Vegetables	29.9	12.0	1.0	9.1					27.7
Fruits	6.4	1.3	.5	2.7					2.1
Total vegetable food	64.9	58.3	2.8	97.6					52.3
Total food	100.0	100.0	100.0	100.0					100.0

TABLE 27.—*Nutrients and potential energy in food purchased in dietary No. 105.*

Kind of food material.	Cost.	Nutrients.			Fuel value.
		Protein.	Fat.	Carbohydrates.	

FOR FAMILY, 10 DAYS.

		Grams.	Grams.	Grams.	Calories.
Food purchased:					
Animal	$1.84	839	5,515	258	55,790
Vegetable	2.02	1,175	160	10,184	48,060
Total	3.86	2,014	5,675	10,442	103,850

PER MAN PER DAY.

Food purchased:					
Animal	.04½	20	134	6	1,360
Vegetable	.05	29	4	249	1,175
Total	.09½	49	138	255	2,535

PERCENTAGES OF TOTAL FOOD PURCHASED.

	Per cent.	Per cent.	Per cent.	Per cent.	Per cent.
Food purchased					
Animal	47.7	41.7	97.2	2.4	53.6
Vegetable	52.3	58.3	2.8	97.6	46.4
Total	100.0	100.0	100.0	100.0	100.0

DIETARY OF A NEGRO FARMER'S FAMILY IN ALABAMA (No. 130).

The study began December 8, 1895, and continued twelve days.
The members of the family and number of meals taken were as follows:

	Meals.
Man 35 years old	36
Woman 35 years old (36 meals × 0.8 meal of man) equivalent to	29
Boy 12 years old (36 meals × 0.6 meal of man) equivalent to	22
Child about 8 years old (36 meals × 0.5 meal of man) equivalent to	18
Child about 6 years old (36 meals × 0.5 meal of man) equivalent to	18
Child about 3 years old (36 meals × 0.4 meal of man) equivalent to	14
Infant (36 meals × 0.3 meal of man) equivalent to	11
Total number of meals	148

Equivalent to one man for forty-nine days.

Remarks.—This was a winter dietary of the same family as No. 100.
The same typical foods were used as in the spring with the addition of
sweet potatoes. The latter were cooked by roasting in hot ashes.

TABLE 28.—*Food materials in dietary No. 130.*

Kind of food material.	Composition.				Weight used.			
	Pro- tein.	Fat.	Carbo- hydrates.	Total cost.	Total food mate- rial.	Nutrients.		
						Pro- tein.	Fat.	Carbo- hydrates.
ANIMAL FOOD.								
Pork:	*Per ct.*	*Per ct.*	*Per ct.*		*Grams.*	*Grams.*	*Grams.*	*Grams.*
Bacon[1]	7.5	66.1		$0.32	1,815	136	1,200	
Lard		100.0		.15	905		905	
Total animal food				.47	2,720	136	2,105	
VEGETABLE FOOD.								
Corn meal[1]	7.5	4.2	65.9	.45	16,330	1,225	686	10,761
Molasses[1]	1.0	.1	71.8	.50	4,990	50	5	3,583
Sweet potatoes	1.5	.6	23.1	.34	20,310	305	122	4,691
Total vegetable food				1.29	41,630	1,580	813	19,035
Total food				1.76	44,350	1,716	2,918	19,035

[1] Average of analyses of similar Alabama foods.

TABLE 29.—*Weights and percentages of food materials and nutritive ingredients used in dietary No. 130.*

Kind of food material.	Food mate- rial.	Nutrients.			Food mate- rial.	Nutrients.			Cost.
		Pro- tein.	Fat.	Carbo- hy- drates.		Pro- tein.	Fat.	Carbo- hy- drates.	
FOR FAMILY, 12 DAYS.	*Grams.*	*Grams.*	*Grams.*	*Grams.*	*Lbs.*	*Lbs.*	*Lbs.*	*Lbs.*	
Pork, lard, etc	2,720	136	2,105		6.00	0.30	4.60		$0.47
Total animal food	2,720	136	2,105		6.00	.30	4.60		.47
Cereals, sugars, starches	21,320	1,275	691	14,344	47.00	2.80	1.50	31.60	.95
Vegetables	20,310	305	122	4,691	44.80	.70	.30	10.40	.34
Total vegetable food	41,630	1,580	813	19,035	91.80	3.50	1.80	42.00	1.29
Total food	44,350	1,716	2,918	19,035	97.80	3.80	6.40	42.00	1.76
PER MAN PER DAY.									
Pork, lard, etc	56	3	43		.12	.01	.09		
Total animal food	56	3	43		.12	.01	.09		.01
Cereals, sugars, starches	435	26	14	293	.96	.06	.03	.65	
Vegetables	414	6	3	96	.91	.01	.01	.21	
Total vegetable food	849	32	17	389	1.87	.07	.04	.86	.02½
Total food	905	35	60	389	1.99	.08	.13	.86	.03½

TABLE 29.—*Weights and percentages of food materials and nutritive ingredients used in dietary No. 130—Continued.*

Kind of food material.	Food material.	Nutrients.				Food material.	Nutrients.			Cost.
		Protein.	Fat.	Carbohydrates.			Protein.	Fat.	Carbohydrates.	
PERCENTAGES OF TOTAL FOOD.	*Per ct.*	*Per ct.*	*Per ct.*	*Per ct.*						*Per ct.*
Pork, lard, etc	6.1	7.9	72.1							26.7
Total animal food ...	6.1	7.9	72.1							26.7
Cereals, sugars, starches..	48.1	74.3	23.7	75.4						54.0
Vegetables	45.8	17.8	4.2	24.6						19.3
Total vegetable food.	93.9	92.1	27.9	100.0						73.3
Total food	100.0	100.0	100.0	100.0						100.0

TABLE 30.—*Nutrients and potential energy in food purchased in dietary No. 130.*

Kind of food material.	Cost.	Nutrients.			Fuel value.
		Protein.	Fat.	Carbohydrates.	
FOR FAMILY, 12 DAYS.		*Grams.*	*Grams.*	*Grams.*	*Calories.*
Food purchased:					
Animal	$0.47	136	2,105		20,130
Vegetable	1.29	1,580	813	19,035	92,080
Total	1.76	1,716	2,918	19,035	112,210
PER MAN PER DAY.					
Food purchased:					
Animal	.01	3	43		410
Vegetable	.02½	32	17	389	1,885
Total	.03½	35	60	389	2,295
PERCENTAGES OF TOTAL FOOD PURCHASED.	*Per cent.*	*Per cent.*	*Per cent.*	*Per cent.*	*Per cent.*
Food purchased:					
Animal	26.7	7.9	72.1		17.9
Vegetable	73.3	92.1	27.9	100.0	82.1
Total	100.0	100.0	100.0	100.0	100.0

DIETARY OF A NEGRO SAWMILL LABORER'S FAMILY IN ALABAMA (No. 131).

The study began December 8, 1895, and continued twelve days (2 meals per day). The members of the family and number of meals taken were as follows:

	Meals.
Man about 26 years old	24
Woman about 25 years old (24 meals × 0.8 meal of man) equivalent to	19
Woman about 18 years old (24 meals × 0.8 meal of man) equivalent to	19
Child about 4 years old (24 meals × 0.4 meal of man) equivalent to..	10
Child about 2 years old (24 meals × 0.3 meal of man) equivalent to..	7
Total number of meals	79

Equivalent to one man for forty days.

Remarks.—This was a winter dietary of the same family as No. 102. The only change in the dietary was in the use of cowpeas. The husband worked as usual in the sawmill while the wife spent her time sewing.

TABLE 31.—*Food materials in dietary No. 13.*

Kind of food material.	Composition.				Weight used.			
	Protein.	Fat.	Carbohydrates.	Total cost.	Total food material.	Nutrients.		
						Protein.	Fat.	Carbohydrates.
ANIMAL FOOD.	Per ct.	Per ct.	Per ct.	Grams.	Grams.	Grams.	Grams.	
Pork:								
Bacon [1]	7.5	66.1	....	$0.72	4,080	306	2,697	....
Lard	......	100.0	....	.19	1,135	......	1,135	....
Butter [1]	1.2	82.4	....	.30	680	8	560	....
Total animal food	......	......	......	1.21	5,895	314	4,392	....
VEGETABLE FOOD.								
Corn meal [1]	7.5	4.2	65.9	.90	3,630	272	153	2,392
Flour [1]	9.6	.9	78.3	.10	13,620	1,308	123	10,664
Sugar (C)	......	......	95.0	.12	1,050	......	......	998
Cow peas	21.6	1.4	61.3	.05	1,815	392	25	1,113
Sweet potatoes	1.5	.6	23.1	.04	2,130	39	13	419
Total vegetable food	......	......	......	1.21	22,245	2,011	314	15,586
Total food	......	......	......	2.42	28,140	2,325	4,706	15,586

[1] Average of analyses of similar Alabama foods.

TABLE 32.—*Weights and percentages of food materials and nutritive ingredients used in Dietary No. 131.*

Kind of food material.	Food material.	Nutrients.			Food material.	Nutrients.			Cost.
		Protein.	Fat.	Carbohydrates.		Protein.	Fat.	Carbohydrates.	
FOR FAMILY, 12 DAYS.	Grams.	Grams.	Grams.	Grams.	Lbs.	Lbs.	Lbs.	Lbs.	
Pork, lard, etc	5,215	306	3,832	......	11.50	0.70	8.50	......	$0.91
Butter	680	8	560	......	1.50	......	1.20	......	.30
Total animal food	5,895	314	4,392	......	13.00	.70	9.70	......	1.21
Cereals, sugars, starches	18,300	1,580	276	14,054	40.30	3.50	.60	31.00	1.12
Vegetables	3,945	431	38	1,532	8.70	.90	.10	3.40	.09
Total vegetable food	22,245	2,011	314	15,586	49.00	4.40	.70	34.40	1.21
Total food	28,140	2,325	4,706	15,586	62.00	5.10	10.40	34.40	2.42
PER MAN PER DAY.									
Pork, lard, etc	130	8	96	......	.28	.02	.21	......	......
Butter	17	......	14	......	.04	......	.03	......	......
Total animal food	147	8	110	......	.32	.02	.24	......	.03
Cereals, sugars, starches	457	39	7	351	1.01	.09	.02	.77	......
Vegetables	99	11	1	39	.22	.02	......	.09	......
Total vegetable food	556	50	8	390	1.23	.11	.02	.86	.03
Total food	703	58	118	390	1.55	.13	.26	.86	.06
PERCENTAGES OF TOTAL FOOD.	Per ct.	Per ct.	Per ct.	Per ct.					Per ct.
Pork, lard, etc	18.5	13.2	81.4	......	......				37.6
Butter	2.4	.3	11.9	......	......				12.4
Total animal food	20.9	13.5	93.3	......	......				50.0
Cereals, sugars, starches	65.1	68.0	5.9	90.2	......				46.3
Vegetables	14.0	18.5	.8	9.8	......				3.7
Total vegetable food	79.1	86.5	6.7	100.0	......				50.0
Total food	100.0	100.0	100.0	100.0	......				100.0

TABLE 33.—*Nutrients and potential energy in food purchased in dietary No. 131.*

Kind of food material.	Cost.	Nutrients.			Fuel value.
		Protein.	Fat.	Carbo- hydrates.	
FOR FAMILY, 12 DAYS.					
Food purchased:		*Grams.*	*Grams.*	*Grams.*	*Calories.*
Animal	$1.21	314	4,392	....	42,135
Vegetable	1.21	2,011	314	15.586	75,065
Total	2.42	2,325	4,706	15.586	117,200
PER MAN PER DAY.					
Food purchased:					
Animal	.03	8	110	....	1,055
Vegetable	.03	50	8	390	1,880
Total	.06	58	118	390	2,935
PERCENTAGES OF TOTAL FOOD PURCHASED.					
Food purchased:	*Per cent.*	*Per cent.*	*Per cent.*	*Per cent.*	*Per cent.*
Animal	50.0	13.5	93.3	....	36.0
Vegetable	50.0	86.5	6.7	100.0	64.0
Total	100.0	100.0	100.0	100.0	100.0

DIETARY OF A NEGRO FARMER'S FAMILY IN ALABAMA (No. 132).

The study began December 8, 1895, and continued twelve days.

The members of the family and number of meals taken were as follows:

	Meals.
Man about 40 years old	36
Woman about 47 years old (36 meals × 0.8 meal of man) equivalent to	29
Girl about 18 years old (36 meals × 0.8 meal of man) equivalent to	29
Child about 11 years old (36 meals × 0.6 meal of man) equivalent to	22
Child about 8 years old (36 meals × 0.5 meal of man) equivalent to	18
Infant (36 meals × 0.3 meal of man) equivalent to	11
Total number of meals	115

Equivalent to one man for forty-eight days.

Remarks.—This family lived in a one-room log cabin on a plantation of 1,800 acres about 5½ miles from Tuskegee. They had a small barn rudely constructed of pine logs. The live stock consisted of two hogs, three hens, and a turkey. They cultivated a "one mule farm" of 30 acres raising cotton almost entirely and that under mortgage. The winter was spent in almost total idleness. The house furniture consisted of two beds, a rude pine table, four homemade chairs, a chest, and a clock worth about $10.

TABLE 34.—*Food materials in dietary No. 132.*

Kind of food material.	Composition.			Total cost.	Total food mate- rial.	Weight used.		
	Pro- tein.	Fat.	Carbo- hydrates.			Nutrients.		
						Pro- tein	Fat.	Carbo- hydrates.
ANIMAL FOOD.								
Pork:	*Per ct.*	*Per ct.*	*Per ct.*		*Grams.*	*Grams.*	*Grams.*	*Grams.*
Fresh pork [1]	10.6	41.9	....	$0.32	2,355	250	971	....
Bacon [1]	7.5	66.1	....	.52	1,815	136	1,200	....
Lard	....	100.0	....	.23	1,360	....	1,360	....
Total animal food	....	....	....	1.07	5,530	386	3,531	....

[1] Average of analyses of similar Alabama foods.

TABLE 34.—*Food materials in dietary No. 13:*—Continued.

Kind of food material.	Composition.			Total cost.	Total food material.	Weights used.		
	Pro-tein.	Fat.	Carbo-hydrates.			Nutrients.		
						Pro-tein.	Fat.	Carbo-hydrates.
VEGETABLE FOOD.	Per ct.	Per ct.	Per ct.		Grams.	Grams.	Grams.	Grams.
Corn meal [1]	7.5	4.2	65.9	$0.25	9,070	680	381	5,977
Molasses [1]	1.0	.1	71.8	.41	4,140	41	4	2,973
Greens [1]	3.8	.9	8.9	.05	905	34	8	81
Sweet potatoes	1.5	.6	23.1	.13	7,710	116	46	1,781
Total vegetable food				.84	21,825	871	439	10,812
Total food				1.91	27,355	1,257	3,970	10,812

[1] Average of analyses of similar Alabama foods.

TABLE 35.—*Weights and percentages of food materials and nutritive ingredients used in dietary No. 13.*

Kind of food material.	Food material.	Nutrients.			Food material.	Nutrients.			Cost.
		Pro-tein.	Fat.	Carbo-hy-drates.		Pro-tein.	Fat.	Carbo-hy-drates.	
FOR FAMILY, 12 DAYS.	Grams.	Grams.	Grams.	Grams.	Lbs.	Lbs.	Lbs.	Lbs.	
Pork, lard, etc	5,530	386	3,531		12.20	0.90	7.80		$1.07
Total animal food	5,530	386	3,531		12.20	.90	7.80		1.07
Cereals, sugars, starches	13,210	721	385	8,950	29.10	1.60	.90	19.70	.66
Vegetables	8,615	150	54	1,862	19.00	.30	.10	4.10	.18
Total vegetable food	21,825	871	439	10,812	48.10	1.90	1.00	23,80	.84
Total food	27,355	1,257	3,970	10,812	60,30	2.80	8.80	23,80	1.91
PER MAN PER DAY.									
Pork, lard, etc	115	8	74		.25	.02	.16		
Total animal food	115	8	74		.25	.02	.16		.02
Cereals, sugars, starches	275	15	8	186	.61	.03	.02	.41	
Vegetables	180	3	1	39	.40	.01		.09	
Total vegetable food	455	18	9	225	1.01	.04	.02	.50	.013
Total food	570	26	83	225	1.26	.06	.18	.50	.04
PERCENTAGES OF TOTAL FOOD.									
Pork, lard, etc	20.2	30.7	88.9						56.0
Total animal food	20.2	30.7	88.9						56.0
Cereals, sugars, starches	48.3	57.4	9.7	82.8					34.6
Vegetables	31.5	11.9	1.4	17.2					9.4
Total vegetable food	79.8	69.3	11.1	100.0					44.0
Total food	100.0	100.0	100.0	100.0					100.0

TABLE 36.—*Nutrients and potential energy in food purchased in dietary No. 13.*

Kind of food material.	Cost.	Nutrients.			Fuel value.
		Protein.	Fat.	Carbo-hydrates.	
FOR FAMILY, 12 DAYS.					
Food purchased:		Grams.	Grams.	Grams.	Calories
Animal	$1.07	386	3,531		34,420
Vegetable	.84	871	439	10,812	51,980
Total	1.91	1,257	3,970	10,812	86,400

TABLE 36.—*Nutrients and potential energy in food purchased in dietary No. 132*—Cont'd.

Kind of food material.	Cost.	Nutrients.			Fuel value.
		Protein.	Fat.	Carbo-hydrates.	
PER MAN PER DAY.					
Food purchased:		*Grams.*	*Grams.*	*Grams.*	*Calories.*
Animal	$0.02¼	8	74	....	720
Vegetable	.01¾	18	9	225	1,080
Total	.04	26	83	225	1,800
PERCENTAGES OF TOTAL FOOD PURCHASED.					
Food purchased:		*Per cent.*	*Per cent.*	*Per cent.*	*Per cent.*	*Per cent.*
Animal	56.0	30.7	88.9	....	39.8
Vegetable	44.0	69.3	11.1	100.0	60.2
Total	100.0	100.0	100.0	100.0	10.00

DIETARY OF A NEGRO (WOMAN) FARMER'S FAMILY IN ALABAMA (No. 133).

The study began December 8, 1895, and continued twelve days (2 meals per day). The members of the family and number of meals taken were as follows:

	Meals.
Woman 60 years old (24 meals × 0.8 meal of man) equivalent to....	19
Woman 32 years old (24 meals × 0.8 meal of man) equivalent to....	19
Girl 12 years old (24 meals × 0.6 meal of man) equivalent to........	14
Boy 11 years old (24 meals × 0.6 meal of man) equivalent to........	14
Boy 7 years old (24 meals × 0.5 meal of man) equivalent to	12
Total number of meals	78

Equivalent to one man for thirty-nine days.

Remarks.—This family lives on a plantation about 7 miles from Tuskegee. The home life and conditions were probably the same as thirty years ago. The house was a one-room frame structure, roughly finished and very simply furnished. The live stock, kept in a barn, consisted of a cow, an ox, and six chickens. The rent of their farm was one and one-half bales of cotton. The family spent most of the winter in splitting pine rails and fishing.

TABLE 37.—*Food materials in dietary No. 133.*

Kind of food material.	Composition.			Total cost.	Total food material.	Weight used.		
	Protein.	Fat.	Carbo-hydrates.			Nutrients.		
						Protein.	Fat.	Carbo-hydrates.
ANIMAL FOOD.	*Per ct.*	*Per ct.*	*Per cent.*		*Grams.*	*Grams.*	*Grams.*	*Grams.*
Beef [1]	19.4	5.2	....	$0.10	455	88	24	....
Bacon [1]	7.5	66.1	....	.72	4,080	306	2,697	....
Lard [1]	....	100.0	....	.23	1,360	......	1,360	....
Total animal food				1.05	5,895	394	4,081	....
VEGETABLE FOOD.								
Corn meal [1]	7.5	4.2	65.9	.57	22,680	1,701	953	14,946
Flour [1]	9.6	.9	78.3	.63	8,620	828	78	6,749
Molasses [1]	1.0	.1	71.8	.50	4,990	50	5	3,583
Greens [1]	3.8	.9	8.9	.03	510	19	5	45
Total vegetable food				1.73	36,800	2,598	1,041	25,323
Total food				2.78	42,695	2,992	5,122	25,323

[1] Average of analyses of similar Alabama foods.

TABLE 38.— *Weights and percentages of food materials and nutritive ingredients used in dietary No. 133.*

Kind of food material.	Food mate- rial.	Nutrients.			Food mate- rial.	Nutrients.			Cost.
		Pro- tein.	Fat.	Carbo- hy- drates.		Pro- tein.	Fat.	Carbo- hy- drates.	
FOR FAMILY, 12 DAYS.									
	Grams.	*Grams.*	*Grams.*	*Grams.*	*Lbs.*	*Lbs.*	*Lbs.*	*Lbs.*	
Beef, veal, and mutton....	455	88	24	....	1.00	0.20	0.10	....	$0.10
Pork, lard, etc	5,440	306	4,057	....	12.00	.70	8.90	....	.95
Total animal food...	5,895	394	4,081	....	13.00	.90	9.00	....	1.05
Cereals, sugars, starches..	36,290	2,579	1,036	25,278	80.00	5.70	2.30	55.70	1.70
Vegetables................	510	19	5	45	1.10	....	....	.10	.03
Total vegetable food.	36,800	2,598	1,041	25,323	81.10	5.70	2.30	55.80	1.73
Total food....	42,695	2,992	5,122	25,323	94.10	6.60	11.30	55.80	2.78
PER MAN PER DAY.									
Beef, veal, and mutton....	12	2	1	....	.03	....	....	....	....
Pork, lard, etc	139	8	104	....	.30	.02	.23	....	....
Total animal food...	151	10	105	....	.33	.02	.23	....	.62¾
Cereals, sugars, starches..	931	66	26	648	2.05	.15	.06	1.43	....
Vegetables................	13	1	....	1	.03	....	....	....	....
Total vegetable food.	944	67	26	649	2.08	.15	.06	1.43	.04¼
Total food	1,095	77	131	649	2.41	.17	.29	1.43	.07
PERCENTAGES OF TOTAL FOOD.									
	Per ct.	*Per ct.*	*Per ct.*	*Per ct.*					*Per ct.*
Beef, veal, and mutton....	1.1	3.0	.5	....	....	....	....	....	3.6
Pork, lard, etc	12.7	10.2	79.2	....	....	....	....	....	34.2
Total animal food...	13.8	13.2	79.7	....	....	....	....	....	37.8
Cereals, sugars, starches..	85.0	86.2	20.2	99.8	....	....	....	....	61.1
Vegetables................	1.2	.6	.1	.2	....	....	....	....	1.1
Total vegetable food.	86.2	86.8	20.3	100.0	....	....	....	....	62.2
Total food..........	100.0	100.0	100.0	100.0	....	....	....	....	100.0

TABLE 39.— *Nutrients and potential energy in food purchased in dietary No. 133.*

Kind of food material.	Cost.	Nutrients.			Fuel value.
		Protein.	Fat.	Carbo- hydrates.	
FOR FAMILY, 12 DAYS.					
		Grams.	*Grams.*	*Grams.*	*Calories.*
Food purchased :					
Animal	$1.05	394	4,081	....	39,570
Vegetable.................	1.73	2,598	1,041	25,323	124,160
Total.....................	2.78	2,992	5,122	25,323	163,730
PER MAN PER DAY.					
Food purchased :					
Animal	.02⅔	10	105	....	1,020
Vegetable.................	.04¼	67	26	649	3,175
Total.....................	.07	77	131	649	4,195
PERCENTAGES OF TOTAL FOOD PURCHASED.					
Food purchased :		*Per cent.*	*Per cent.*	*Per cent.*	*Per cent.*
Animal	37.8	13.2	79.7	....	24.2
Vegetable.................	62.2	86.8	20.3	100.0	75.8
Total.....................	100.0	100.0	100.0	100.0	100.0

DIETARY OF A NEGRO FARMER'S FAMILY IN ALABAMA (No. 134).

The study began January 6, 1896, and continued eleven days (2 meals per day).
The members of the family and number of meals taken were as follows:

	Meals.
Man 46 years old	22
Woman 40 years old (22 meals × 0.8 meal of man) equivalent to	18
Girl 20 years old (22 meals × 0.8 meal of man) equivalent to	18
Total number of meals	58

Equivalent to one man for twenty-nine days.

Remarks.—This family was located on a large plantation, 8½ miles from
Tuskegee, and occupied one room of a two-room frame cabin, the other
half being occupied by another family. They managed a one-mule farm
of 30 acres, which was free from mortgage. The husband worked in
the swamp during the winter cutting fence rails. In the summer the
family had a small garden with a general assortment of vegetables.
The live stock consisted of a mule and a hog.

TABLE 40.—*Food materials in dietary No. 134.*

Kind of food material.	Composition.			Total cost.	Total food material.	Weight used.		
	Protein.	Fat.	Carbohydrates.			Nutrients.		
						Protein.	Fat.	Carbohydrates.
ANIMAL FOOD.	*Per ct.*	*Per ct.*	*Per cent.*		*Grams.*	*Grams.*	*Grams.*	*Grams.*
Pork:								
Fresh pork[1]	10.6	41.0	....	$1.00	4,535	481	1,900	....
Bacon[1]	7.5	66.1	....	.16	905	68	598	....
Lard		100.0	....	.14	820	....	820	....
Total animal food				1.30	6,260	549	3,318	....
VEGETABLE FOOD.								
Corn meal[1]	7.5	4.2	65.9	.45	16,330	1,225	686	10,761
Flour[1]	9.6	.9	78.3	.45	6,805	653	61	5,329
Molasses[1]	1.0	.1	71.8	.16	1,560	16	1	1,120
Sweet potatoes	1.5	.6	23.1	.07	4,160	62	25	961
Total vegetable food				1.13	28,855	1,956	773	18,171
Total food				2.43	35,115	2,505	4,091	18,171

[1] Average of analyses of similar Alabama foods.

TABLE 41.—*Weights and percentages of food materials and nutritive ingredients used in dietary No. 134.*

Kind of food material.	Food material.	Nutrients.			Food material.	Nutrients.			Cost.
		Protein.	Fat.	Carbohydrates.		Protein.	Fat.	Carbohydrates.	
FOR FAMILY, 11 DAYS.	*Grams.*	*Grams.*	*Grams.*	*Grams.*	*Lbs.*	*Lbs.*	*Lbs.*	*Lbs.*	
Pork, lard, etc	6,260	549	3,318	....	13.80	1.20	7.30	....	$1.30
Total animal food	6,260	549	3,318	....	13.80	1.20	7.30	....	1.30
Cereals, sugars, starches	24,695	1,894	748	17,210	54.40	4.20	1.60	38.00	1.06
Vegetables	4,160	62	25	961	9.20	.10	.10	2.10	07
Total vegetable food	28,855	1,956	773	18,171	63.60	4.30	1.70	40.10	1.13
Total food	35,115	2,505	4,091	18,171	77.40	5.50	9.00	40.10	2.43

TABLE 41.—*Weights and percentages of food materials and nutritive ingredients used in dietary No. 134*—Continued.

Kind of food material.	Food mate- rial.	Nutrients.			Food mate- rial.	Nutrients.			Cost.
		Pro- tein.	Fat.	Carbo- hy- drates.		Pro- tein.	Fat.	Carbo- hy- drates.	
PER MAN PER DAY.	*Grams.*	*Grams.*	*Grams.*	*Grams.*	*Lbs.*	*Lbs.*	*Lbs.*	*Lbs.*	
Pork, lard, etc	216	19	114	....	0.48	0.04	0.25	....	....
Total animal food...	216	19	114	....	.48	.04	.25	....	$0.04½
Cereals, sugars, starches...	852	65	26	594	1.88	.14	.06	1.31	....
Vegetables	143	2	1	33	.31	.01	....	.07	....
Total vegetable food.	995	67	27	627	2.19	.15	.06	1.38	.04
Total food	1,211	86	141	627	2.67	.19	.31	1.38	.08½
PERCENTAGES OF TOTAL FOOD.	*Per ct.*	*Per ct.*	*Per ct.*	*Per ct.*					*Per ct.*
Pork, lard, etc	17.8	21.9	81.1	....					53.5
Total animal food...	17.8	21.9	81.1	....					53.5
Cereals, sugars, starches..	70.3	75.6	18.3	94.7					43.6
Vegetables	11.9	2.5	.6	5.3					2.9
Total vegetable food.	82.2	78.1	18.9	100.0					46.5
Total food	100.0	100.0	100.0	100.0					100.0

TABLE 42.—*Nutrients and potential energy in food purchased in dietary No. 134.*

Kind of food material.	Cost.	Nutrients.			Fuel value.
		Protein.	Fat.	Carbo- hydrates.	
FOR FAMILY, 11 DAYS.		*Grams.*	*Grams.*	*Grams.*	*Calories.*
Food purchased:					
Animal	$1.30	549	3,318	....	33,110
Vegetable	1.13	1,956	773	18,171	89,710
Total	2.43	2,505	4,091	18,171	122.820
PER MAN PER DAY.					
Food purchased:					
Animal	.04½	19	114	....	1,140
Vegetable	.04	67	27	627	3,095
Total	.08½	86	141	627	4,235
PERCENTAGES OF TOTAL FOOD PURCHASED.		*Per cent.*	*Per cent.*	*Per cent.*	*Per cent.*
Food purchased:					
Animal		53.5	17.8	21.9	81.1	27.0
Vegetable		46.5	82.2	78.1	18.9	73.0
Total		100.0	100.0	100.0	100.0	100.0

DIETARY OF A NEGRO FARMER'S FAMILY IN ALABAMA (No. 135).

The study began December 12, 1895, and continued fifteen days (2 meals per day). The members of the family and number of meals taken were as follows:

	Meals.
Woman 60 years old (30 meals × 0.8 meal of man) equivalent to	24
Man 25 years old	30
Total number of meals	54

Equivalent to one man for twenty-seven days.

12246—No. 38——4

Remarks.—This family lived on a plantation of some 1,400 acres. The mother hired about 30 acres, paying a bale and a half of cotton, and actively assisted her son in plowing and caring for the farm during the working season. The son occasionally hauled a small load of pine wood to the village in the winter and the mother passed the time as best she could in one of the two rooms in their cabin. The live stock consisted of two oxen, three hogs, and a few chickens.

TABLE 43.—*Food materials in dietary No. 135.*

Kind of food material.	Composition.				Weight used.			
	Pro-tein.	Fat.	Carbo-hydrates.	Total cost.	Total food mate-rial.	Nutrients.		
						Pro-tein.	Fat.	Carbo-hydrates.
ANIMAL FOOD.	*Per ct.*	*Per ct.*	*Per cent.*		*Grams.*	*Grams.*	*Grams.*	*Grams.*
Pork, bacon[1]	7.5	66.1	....	$0.45	2,550	191	1,686	....
VEGETABLE FOOD.								
Corn meal[1]	7.5	4.2	65.9	.15	13,635	1,023	573	8,985
Flour[1]	9.6	.9	78.3	.38	2,270	218	21	1,777
Molasses[1]	1.0	.1	71.8	.17	1,700	17	2	1,220
Sugar	......	......	100.0	.06	395	......	......	395
Total vegetable food				.76	18,000	1,258	596	12,377
Total food				1.21	20,550	1,449	2,282	12,377

[1] Average of analyses of similar Alabama foods.

TABLE 44.—*Weights and percentages of food materials and nutritive ingredients used in dietary No. 135.*

Kind of food material.	Food ma-terial.	Nutrients.			Food ma-terial.	Nutrients.			Cost.
		Pro-tein.	Fat.	Carbo-hy-drates.		Pro-tein.	Fat.	Carbo-hy-drates.	
FOR FAMILY, 15 DAYS.	*Grams.*	*Grams.*	*Grams.*	*Grams.*	*Lbs.*	*Lbs.*	*Lbs.*	*Lbs.*	
Pork, lard, etc	2,550	191	1,686	......	5.60	0.40	3.70	......	$0.45
Cereals, sugars, starches	18,000	1,258	596	12,377	39.70	2.80	1.30	27.30	.76
Total food	20,550	1,449	2,282	12,377	45.30	3.20	5.00	27.30	1.21
PER MAN PER DAY.									
Pork, lard, etc	91	7	63	......	.21	.02	14	......	.01¾
Cereals, sugars, starches	667	47	22	458	1.47	.10	.19	1.01	.02¾
Total food	761	54	85	458	1.68	.12	.33	1.01	.04½
PERCENTAGES OF TOTAL FOOD.	*Per ct.*	*Per ct.*	*Per ct.*	*Per ct.*					*Per ct*
Pork, lard, etc	12.4	13.2	73.9	......	100.0				37.2
Cereals, sugars, starches	87.6	86.8	26.1	100.0					62.8
Total food	100.0	100.0	100.0	100.0					100.0

TABLE 45.—*Nutrients and potential energy in food purchased in dietary No. 135.*

Kind of food material.	Cost.	Protein.	Fat.	Carbo-hydrates.	Fuel value.

FOR FAMILY, 15 DAYS.

		Grams.	Grams.	Grams.	Calories.
Food purchased:					
Animal	$0.45	191	1,686	....	16,465
Vegetable	.76	1,258	596	12,377	61,445
Total	1.21	1,449	2,282	12,377	77,910

PER MAN PER DAY.

Food purchased:					
Animal	.01¼	7	63	....	615
Vegetable	.02¼	47	22	458	2,275
Total	.04½	54	85	458	2,890

PERCENTAGES OF TOTAL FOOD PURCHASED.

Food purchased:		Per cent.	Per cent.	Per cent.	Per cent.	Per cent.
Animal		37.2	13.2	73.9	....	21.1
Vegetable		62.8	86.8	26.1	100.0	78.9
Total		100.0	100.0	100.0	100.0	100.0

DIETARY OF A NEGRO FARMER'S FAMILY IN ALABAMA (No. 136).

The study began January 6, 1896, and continued eighteen days (2 meals per day). The members of the family and number of meals taken were as follows:

	Meals.
Man 48 years old	36
Man 50 years old	36
Woman 32 years old (36 meals × 0.8 meal of man) equivalent to	29
Girl 13 years old (36 meals × 0.6 meal of man) equivalent to	21
Total number of meals	122

Equivalent to one man for sixty-one days.

Remarks.—The family carried on a 40-acre farm on one of the oldest plantations in the Tuskegee section, some sixteen negro families living on the same place. The crop of 1896, entirely of cotton, was mortgaged in advance to secure winter provisions and farm implements. A small corncrib of logs and a barn similarly constructed constituted the out-buildings. The husband and wife spent a part of the winter days in cutting shingles in a neighboring swamp. The live stock consisted of a mule, three pigs, and a half dozen chickens. They occupied one room of a two-room cottage and rented the other.

TABLE 46.—*Food materials in dietary No. 136.*

Kind of food material.	Composition.			Total cost.	Weight used.			
	Pro-tein.	Fat.	Carbo-hydrates.		Total food mate-rial.	Nutrients.		
						Pro-tein.	Fat.	Carbo-hydrates.
ANIMAL FOOD.	Per ct.	Per ct.	Per cent.		Grams.	Grams.	Grams.	Grams.
Pork, bacon[1]	7.5	66.1	....	$1.70	9,610	721	6,352	....
VEGETABLE FOOD.								
Corn meal[1]	7.5	4.2	65.9	.45	20,640	1,548	867	13,602
Flour[1]	9.6	.9	78.3	.57	6,745	648	61	5,281
Molasses[1]	1.0	.1	71.8	.42	4,215	42	4	3,027
Sugar	....	....	100.0	.02	115	....	....	115
Greens[1]	2.8	.9	8.9	.04	735	28	7	65
Total vegetable food				1.50	32,450	2,266	939	22,090
Total food				3.20	42,060	2,987	7,291	22,090

[1]Average of analyses of similar Alabama foods.

52

TABLE 17.—*Weights and percentages of food materials and nutritive ingredients used in dietary No. 136.*

Kind of food material.	Food material.	Nutrients.			Food material.	Nutrients.			Cost.
		Protein.	Fat.	Carbohydrates.		Protein.	Fat.	Carbohydrates.	
FOR FAMILY, 18 DAYS.	*Grams.*	*Grams.*	*Grams.*	*Grams.*	*Lbs.*	*Lbs.*	*Lbs.*	*Lbs.*	
Pork, lard, etc	9,610	721	6,352		21.20	1.60	14.00		$1.70
Total animal food....	9,610	721	6,352		21.20	1.60	14.00		1.70
Cereals, sugars, starches ..	31,715	2,238	932	22,025	69.90	4.90	2.10	48.60	1.46
Vegetables	735	28	7	65	1.60	.10		.10	.04
Total vegetable food.	32,450	2,266	939	22,090	71.50	5.00	2.10	48.70	1.50
Total food	42,060	2,987	7,291	22,090	92.70	6.60	16.10	48.70	3.20
PER MAN PER DAY.									
Pork, lard, etc	158	12	104		.35	.03	.23		
Total animal food...	158	12	104		.35	.03	.23		.02¾
Cereals, sugars, starches ..	520	37	15	361	1.14	.08	.03	.80	
Vegetables	12			1	.03				
Total vegetable food.	532	37	15	362	1.17	.08	.03	.80	.02½
Total food	630	49	119	362	1.52	.11	.26	.80	.05¼
PERCENTAGES OF TOTAL FOOD.	*Per ct.*	*Per ct.*	*Per ct.*	*Per ct.*					*Per ct.*
Pork, lard, etc	22.8	24.1	87.1						53.1
Total animal food...	22.8	24.1	87.1						53.1
Cereals, sugars, starches ..	75.4	74.9	12.8	99.7					45.6
Vegetables	1.8	1.0	.1	.3					1.3
Total vegetable food.	77.2	75.9	12.9	100.0					46.9
Total food	100.0	100.0	100.0	100.0					100.0

TABLE 18.—*Nutrients and potential energy in food purchased in dietary No. 136.*

Kind of food material.	Cost.	Nutrients.			Fuel value.
		Protein.	Fat.	Carbohydrates.	
FOR FAMILY, 18 DAYS.		*Grams.*	*Grams.*	*Grams.*	*Calories.*
Food purchased:					
Animal	$1.70	721	6.352		62,030
Vegetable	1.50	2,266	939	22,090	108,590
Total	3.20	2,987	7,291	22,090	170,620
PER MAN PER DAY.					
Food purchased:					
Animal	.02¾	12	104		1,015
Vegetable	.02½	37	15	362	1,775
Total	.05¼	49	119	362	2,790
PERCENTAGES OF TOTAL FOOD PURCHASED.					
Food purchased:	*Per cent.*	*Per cent.*	*Per cent.*	*Per cent.*	*Per cent.*
Animal	53.1	24.1	87.1		36.4
Vegetable	46.9	75.9	12.9	100.0	63.6
Total	100.0	100.0	100.0	100.0	100.0

DIETARY OF A NEGRO FARMER'S FAMILY IN ALABAMA (No. 137).

The study began January 28, 1896, and continued eighteen days.
The members of the family and number of meals taken were as follows:

	Meals.
Man 57 years old	54
Girl 9 years old (54 meals × 0.5 meal of man) equivalent to	27
Girl 7 years old (54 meals × 0.5 meal of man) equivalent to	27
Girl 5 years old (54 meals × 0.4 meal of man) equivalent to	22
Boy 3 years old (54 meals × 0.4 meal of man) equivalent to	21
Infant 1 year old (54 meals × 0.3 meal of man) equivalent to	16
Total number of meals	167

Equivalent to one man for fifty-six days.

Remarks.—This family lived in one room of a two-room log cabin on
a plantation 7 miles from Tuskegee. They were miserably poor, sub-
sisting for days at a time on nothing but corn pone. The man managed
a one-mule farm, raised only cotton, and paid a bale and a half of the
same for rent. The live stock consisted of two chickens, a mule being
hired during the cotton season. During the winter they made chairs
for sale among the neighbors.

TABLE 49.—*Food materials in dietary No. 137.*

Kind of food material.	Composition.			Total cost	Weight used.			
	Protein.	Fat.	Carbohydrates.		Total food material.	Nutrients.		
						Protein.	Fat.	Carbohydrates.
ANIMAL FOOD.	*Per ct.*	*Per ct.*	*Per cent.*		*Grams.*	*Grams.*	*Grams.*	*Grams.*
Pork, bacon [1]	7.5	66.1		$0.18	1,040	78	687	
Total animal food				.18	1,040	78	687	
VEGETABLE FOOD.								
Corn meal [1]	7.5	4.2	65.9	.11	19,405	1,455	815	12,786
Flour [1]	9.6	.9	78.3	.54	1,730	166	16	1,354
Molasses [1]	1.0	.1	71.8	.40	3,995	40	4	2,868
Total vegetable food				1.05	25,130	1,661	835	17,010
Total food				1.23	26,170	1,739	1,522	17,010

[1] Average of analyses of similar Alabama foods.

TABLE 50.—*Weights and percentages of food materials and nutritive ingredients used in dietary No. 137.*

Kind of food material.	Food material.	Nutrients.			Food material.	Nutrients.			Cost.
		Protein.	Fat.	Carbohydrates.		Protein.	Fat.	Carbohydrates.	
FOR FAMILY, 18 DAYS.	*Grams.*	*Grams.*	*Grams.*	*Grams.*	*Lbs.*	*Lbs.*	*Lbs.*	*Lbs.*	
Pork, lard, etc	1,040	78	687		2.30	0.20	1.50		$0.18
Cereals, sugars, starches	25,130	1,661	835	17,010	55.40	3.60	1.80	37.50	1.05
Total food	26,170	1,739	1,522	17,010	57.70	3.80	3.30	37.50	1.23

54

TABLE 50.—*Weights and percentages of food materials and nutritive ingredients used in dietary No. 137—Continued.*

Kind of food material.	Food material.	Nutrients.			Food material.	Nutrients.			Cost.
		Protein.	Fat.	Carbohydrates.		Protein.	Fat.	Carbohydrates.	
PER MAN PER DAY.	Grams.	Grams.	Grams.	Grams.	Lbs.	Lbs.	Lbs.	Lbs.	
Pork, lard, etc	18	1	12	....	0.04	....	0.03	....	$0.00¼
Cereals, sugars, starches..	449	30	15	304	.99	0.07	.03	0.67	.02
Total food	467	31	27	304	1.03	.07	.06	.67	.02¼
PERCENTAGES OF TOTAL FOOD.	Per ct.	Per ct.	Per ct.	Per ct.					Per ct.
Pork, lard, etc	4.0	4.5	45.1	....					14.6
Cereals, sugars, starches..	96.0	95.5	54.9	100.0					85.4
Total food	100.0	100.0	100.0	100.0					100.0

TABLE 51.—*Nutrients and potential energy in food purchased in dietary No. 137.*

Kind of food material.	Cost.	Nutrients.			Fuel value.
		Protein.	Fat.	Carbohydrates.	
FOR FAMILY, 18 DAYS.		Grams.	Grams.	Grams.	Calories.
Food purchased:					
Animal	$0.18	78	687	....	6,710
Vegetable	1.05	1,661	835	17,010	84,320
Total	1.23	1,739	1,522	17,010	91,030
PER MAN PER DAY.					
Food purchased:					
Animal	.00¼	1	12	....	115
Vegetable	.02	30	15	304	1,510
Total	.02¼	31	27	304	1,625
PERCENTAGES OF TOTAL FOOD PURCHASED.					
Food purchased:		Per cent.	Per cent.	Per cent.	Per cent.
Animal	14.6	4.5	45.1	100.0	7.4
Vegetable	85.4	95.5	54.9	100.0	92.6
Total	100.0	100.0	100.0	100.0	100.0

DIETARY OF A NEGRO FARMER'S FAMILY IN ALABAMA (No. 138).

The study began January 28, 1896, and continued sixteen days (2 meals per day). The members of the family and number of meals taken were as follows:

	Meals.
Woman 28 years old (32 meals × 0.8 meal of man) equivalent to....	26
Boy 12 years old (32 meals × 0.6 meal of man) equivalent to	19
Total number of meals	45

Equivalent to one man for twenty-two days.

Remarks.—This family managed a two-mule farm on a large plantation 8 miles from Tuskegee, paying 2½ bales of cotton as rent. The one-room cabin was neatly furnished, and the woman in spare moments made dresses and bonnets for her neighbors. She was the owner of a sewing machine. The property of the family was this year free from mortgage. The live stock consisted of two mules, one cow, and a few hens. They had a corn crib and small barn made of logs.

TABLE 52.—*Food materials in dietary No. 138.*

Kind of food material.	Composition.				Weight used.			
	Protein.	Fat.	Carbohydrates.	Total cost.	Total food material.	Nutrients.		
						Protein.	Fat.	Carbohydrates.
ANIMAL FOOD.	*Per ct.*	*Per ct.*	*Per cent.*		*Grams.*	*Grams.*	*Grams.*	*Grams.*
Pork, bacon [1]	7.5	66.1		$0.55	3,095	232	2,046	
Butter [1]	1.2	82.4		.13	285	3	235	
Milk	3.3	4.0	5.0	.12	1,360	45	54	68
Buttermilk	3.0	.5	4.8	.07	2,720	81	14	131
Total animal food				.87	7,460	361	2,349	199
VEGETABLE FOOD.								
Corn meal [1]	7.5	4.2	65.9	.22	5,725	429	240	3,773
Flour [1]	9.6	.9	78.3	.16	3,290	316	30	2,576
Molasses [1]	1.0	.1	71.8	.40	4,080	41	4	2,929
Sugar (C)			95.0	.09	850			608
Total vegetable food				.87	13,945	786	274	10,086
Total food				1.74	21,405	1,147	2,623	10,285

[1] Average of analyses of similar Alabama foods.

TABLE 53.—*Weights and percentages of food materials and nutritive ingredients used in dietary No. 138.*

Kind of food material.	Food material.	Nutrients.			Food material.	Nutrients.			Cost.
		Protein.	Fat.	Carbohydrates.		Protein.	Fat.	Carbohydrates.	
FOR FAMILY, 16 DAYS.	*Grams.*	*Grams.*	*Grams.*	*Grams.*	*Lbs.*	*Lbs.*	*Lbs.*	*Lbs.*	
Pork, lard, etc	3,095	232	2,046		6.80	0.50	4.50		$0.55
Butter	285	3	235		.60		.50		.13
Milk	1,360	45	54	68	3.00	.10	.10	.20	.12
Buttermilk	2,720	81	14	131	6.00	.20	.10	.30	.07
Total animal food	7,460	361	2,349	199	16.40	.80	5.20	.50	.87
Cereals, sugars, starches	13,945	786	274	10,086	30.80	1.70	.60	22.20	.87
Total vegetable food	13,945	786	274	10,086	30.80	1.70	.60	22.20	.87
Total food	21,405	1,147	2,623	10,285	47.20	2.50	5.80	22.70	1.74
PER MAN PER DAY.									
Pork, lard, etc	141	10	93		.31	.02	.21		
Butter	13		11		.03		.03		
Milk	62	2	2	3	.14			.01	
Buttermilk	123	4	1	6	.27	.01		.01	
Total animal food	339	16	107	9	.75	.03	.24	.02	.04
Cereals, sugars, starches	634	36	13	458	1.40	.08	.03	1.01	
Total vegetable food	634	36	13	458	1.40	.08	.03	1.01	.04
Total food	973	52	120	467	2.15	.11	.27	1.03	.08
PERCENTAGES OF TOTAL FOOD.	*Per ct.*	*Per ct.*	*Per ct.*	*Per ct.*					*Per ct.*
Pork, lard, etc	14.5	20.2	78.0						31.6
Butter	1.3	.3	9.0						7.5
Milk	6.1	3.9	2.1	.6					6.9
Buttermilk	12.7	7.1	.5	1.3					4.0
Total animal food	34.9	31.5	89.6	1.9					50.0
Cereals, sugars, starches	65.1	68.5	10.4	98.1					50.0
Total vegetable food	65.1	68.5	10.4	98.1					50.0
Total food	100.0	100.0	100.0	100.0					100.0

56

TABLE 54.—*Nutrients and potential energy in food purchased in dietary No. 138.*

Kind of food material.	Cost.	Nutrients.			Fuel value.
		Protein.	Fat.	Carbo-hydrates.	
FOR FAMILY, 16 DAYS.					
Food purchased:		Grams.	Grams.	Grams.	Calories.
Animal	$0.87	361	2,349	199	24,140
Vegetable	.87	786	274	10,086	47,125
Total	1.74	1,147	2,623	10,285	71,265
PER MAN PER DAY.					
Food purchased:					
Animal	.04	16	107	9	1,100
Vegetable	.04	36	13	458	2,145
Total	.08	52	120	467	3,245
PERCENTAGES OF TOTAL FOOD PURCHASED.					
Food purchased:	Per cent.	Per cent.	Per cent.	Per cent.	Per cent.
Animal	50.0	31.5	89.6	1.9	33.9
Vegetable	50.0	68.5	10.4	98.1	66.1
Total	100.0	100.0	100.0	100.0	100.0

DIETARY OF A NEGRO FARMER'S FAMILY IN ALABAMA (No. 139).

The study began January 28, 1896, and continued sixteen days.
The members of the family and number of meals taken were as follows:

	Meal.
Man about 33 years old	48
Woman about 30 years old (48 meals × 0.8 meal of man) equivalent to	38
Total number of meals	86

Equivalent to one man for twenty-nine days.

Remarks.—This family lived on a 25-acre farm on a plantation of 1,500 acres, some 7½ miles from Tuskegee. The one-room log cabin was attractively surrounded by flowers and a vegetable garden, and the interior showed some attempts at adornment. They raised cotton, corn, and potatoes; cured their own bacon this year, and made some 15 pounds of lard. The live stock consisted of a mule, four small pigs, and six hens, which were kept in a log barn.

TABLE 55.—*Food materials in dietary No. 139.*

Kind of food material.	Composition.			Total cost.	Total food material.	Weight used.		
	Protein.	Fat.	Carbo-hydrates.			Nutrients.		
						Protein.	Fat.	Carbo-hydrates.
ANIMAL FOOD.								
Pork:	Per ct.	Per ct.	Per cent.		Grams.	Grams.	Grams.	Grams.
Fresh pork[1]	10.6	41.9		$1.04	2,270	241	951	
Bacon[1]	7.5	66.1		.50	5,895	442	3,897	
Lard		100.0		.38	2,270		2,270	
Milk	3.3	4.0	5.0	.09	1,020	33	41	51
Total animal food				2.01	11,455	716	7,159	5
VEGETABLE FOOD.								
Corn meal[1]	7.5	4.2	65.9	.35	13,610	1,021	572	8,968
Flour[1]	9.6	.9	78.3	.38	5,245	504	47	4,107
Rice	7.8	.4	79.0	.07	595	46	2	471
Sugar (C)			95.0	.11	1,030			979
Sweet potatoes	1.5	.6	23.1	.03	1,985	30	12	458
Total vegetable food				.94	22,465	1,601	633	14,983
Total food				2.95	33,920	2,317	7,792	15,034

[1] Average of analyses of similar Alabama foods.

57

TABLE 56.—*Weights and percentages of food materials and nutritive ingredients used in dietary No. 139.*

Kind of food material.	Food material.	Nutrients. Protein.	Fat.	Carbo-hydrates.	Food material.	Nutrients. Protein.	Fat.	Carbo-hydrates.	Cost.
FOR FAMILY, 16 DAYS.	*Grams.*	*Grams.*	*Grams.*	*Grams.*	*Lbs.*	*Lbs.*	*Lbs.*	*Lbs.*	
Pork, lard, etc	10,435	683	7,118	....	23.00	1.50	15.70	....	$1.92
Milk	1,020	33	41	51	2.30	.10	.10	0.10	.09
Total animal food ...	11,455	716	7,159	51	25.30	1.60	15.80	.10	2.01
Cereals, sugars, starches..	20,480	1,571	621	14,525	45.10	3.40	1.40	32.00	.91
Vegetables	1,985	30	12	458	4.40	.10	....	1.00	.03
Total vegetable food.	22,465	1,601	633	14,983	49.50	3.50	1.40	33.00	.94
Total food	33,920	2,317	7,792	15,034	74.80	5.10	17.20	33.10	2.95
PER MAN PER DAY.									
Pork, lard, etc	360	24	246	....	.79	.06	.54	....	....
Milk	35	1	1	1	.08	....	....	....	....
Total animal food...	395	25	247	1	.87	.06	.54	....	.07
Cereals, sugars, starches..	706	54	22	501	1.56	.12	.05	1.10	....
Vegetables	69	1	....	16	.15	....	....	.04	....
Total vegetable food.	775	55	22	517	1.71	.12	.05	1.14	.03¼
Total food	1,170	80	269	518	2.58	.18	.59	1.14	.10¼
PERCENTAGES OF TOTAL FOOD.	*Per ct.*	*Per ct.*	*Per ct.*	*Per ct.*					*Per ct.*
Pork, lard, etc	30.8	29.5	91.4	....					65.1
Milk	3.0	1.4	.5	.3					3.0
Total animal food ...	33.8	30.9	91.9	.3					68.1
Cereals, sugars, starches..	60.4	67.8	8.0	96.6					30.9
Vegetables	5.8	1.3	.1	3.1					1.0
Total vegetable food.	66.2	69.1	8.1	99.7					31.9
Total food	100.0	100.0	100.0	100.0					100.0

TABLE 57.—*Nutrients and potential energy in food purchased in dietary No. 139.*

Kind of food material.	Cost.	Nutrients. Protein.	Fat.	Carbo-hydrates.	Fuel value.
FOR FAMILY, 16 DAYS.		*Grams.*	*Grams.*	*Grams.*	*Calories.*
Food purchased: Animal	$2.01	716	7,159	51	69,720
Vegetable	.94	1,601	633	14,983	73,880
Total	2.95	2,317	7,792	15,034	143,600
PER MAN PER DAY.					
Food purchased: Animal	.07	25	247	1	2,405
Vegetable	.03¼	55	22	517	2,550
Total	.10¼	80	269	518	4,955
PERCENTAGES OF TOTAL FOOD PURCHASED.		*Per cent.*	*Per cent.*	*Per cent.*	*Per cent.*
Food purchased: Animal	68.1	30.9	91.9	0.3	48.6
Vegetable	31.9	69.1	8.1	99.7	51.4
Total	100.0	100.0	100.0	100.0	100.0

DIETARY OF A PLANTATION HAND'S FAMILY IN ALABAMA (No. 140).

The study began January 29, 1896, and continued fifteen days (2 meals per day). The members of the family and number of meals taken were as follows:

	Meals.
Woman 29 years old (30 meals × 0.8 meal of man) equivalent to....	24
Boy 10 years old (30 meals × 0.6 meal of man) equivalent to........	18
Boy 8 years old (30 meals × 0.5 meal of man) equivalent to........	15
Boy 5 years old (30 meals × 0.4 meal of man) equivalent to........	12
Child 2 years old (30 meals × 0.4 meal of man) equivalent to........	12
Total number of meals	81

Equivalent to one man for forty days.

Remarks.—This family lived in a one-room log cabin on a plantation about 7 miles from Tuskegee, and consisted of a mother and four children. The support of the entire family rested on the mother, who worked as a common plantation laborer—plowing and doing the same work as men. There is a barn and an old log shed. The live stock consisted of two hens.

TABLE 58.—*Food materials in dietary No. 140.*

Kind of food material.	Composition.			Total cost.	Weight used.			
	Protein.	Fat.	Carbohydrates.		Total food material.	Nutrients.		
						Protein.	Fat.	Carbohydrates.
ANIMAL FOOD.	*Per ct.*	*Per ct.*	*Per ct.*		*Grams.*	*Grams.*	*Grams.*	*Grams.*
Pork:								
Fresh pork [1]	10.6	41.9	....	$1.66	795	84	333	....
Bacon [1]	7.5	66.1	....	.18	9,440	708	6,240	....
Buttermilk	3.0	.5	4.8	.01	480	14	3	23
Total animal food				1.85	10,715	806	6,576	23
VEGETABLE FOOD.								
Corn meal [1]	7.5	4.2	65.9	.46	16,785	1,259	705	11,061
Sugar (C)	....	....	95.0	.22	1,955	....	....	1,857
Total vegetable food				.68	18,740	1,259	705	12,918
Total food				2.53	29,455	2,065	7,281	12,941

[1] Average of analyses of similar Alabama foods.

TABLE 59.—*Weights and percentages of food materials and nutritive ingredients used in dietary No. 140.*

TABLE 59.—*Weights and percentages of food materials and nutritive ingredients used in dietary No. 140—Continued.*

Kind of food material.	Food material.	Nutrients.			Food material.	Nutrients.			Cost.
		Protein.	Fat.	Carbohydrates.		Protein.	Fat.	Carbohydrates.	
PER MAN PER DAY.	*Grams.*	*Grams.*	*Grams.*	*Grams.*	*Lbs.*	*Lbs.*	*Lbs.*	*Lbs.*	
Pork, lard, etc............	256	20	164	....	0.56	0.04	0.36	....	
Buttermilk.............	12	....	....	1	.03	....	....	....	
Total animal food ...	268	20	164	1	.59	.04	.36	....	$0.04½
Cereals, sugars, starches .	468	32	18	323	1.03	.07	.04	0.71	....
Total vegetable food.	468	32	18	323	1.03	.07	.04	.71	.01¾
Total food...........	736	52	182	324	1.62	.11	.40	.71	.06¼
PERCENTAGES OF TOTAL FOOD.	*Per ct.*	*Per ct.*	*Per ct.*	*Per ct.*					*Per ct.*
Pork, lard, etc............	34.8	38.3	90.3	....	....	....	....	....	72.7
Buttermilk.............	1.6	.7	....	.2	....	....	....	....	.4
Total animal food ...	36.4	39.0	90.3	.2	....	....	....	....	73.1
Cereals, sugars, starches..	63.6	61.0	9.7	99.8	....	....	....	....	26.9
Total vegetable food.	63.6	61.0	9.7	99.8	....	....	....	....	26.9
Total food...........	100.0	100.0	100.0	100.0	....	....	....	....	100.0

TABLE 60.—*Nutrients and potential energy in food purchased in dietary No. 140.*

Kind of food material.	Cost.	Nutrients.			Fuel value.
		Protein.	Fat.	Carbohydrates.	
FOR FAMILY, 15 DAYS.		*Grams.*	*Grams.*	*Grams.*	*Calories.*
Food purchased:					
Animal...................................	$1.85	806	6,576	23	64,560
Vegetable..........................	.68	1,259	705	12,918	64,680
Total...........................	2.53	2,065	7,281	12,941	129,240
PER MAN PER DAY.					
Food purchased:					
Animal...	.04½	20	164	1	1,610
Vegetable..........................	.01¾	32	18	323	1,625
Total...........................	.06¼	52	182	324	3,235
PERCENTAGES OF TOTAL FOOD PURCHASED.		*Per cent.*	*Per cent.*	*Per cent.*	*Per cent.*
Food purchased:					
Animal.................................	73.1	39.0	90.3	.2	50.0
Vegetable..........................	26.9	61.0	9.7	99.8	50.0
Total...........................	100.0	100.0	100.0	100.0	100.0

DIETARY OF A NEGRO FARMER'S FAMILY IN ALABAMA (No. 141).

The study began February 1, 1896, and continued fifteen days.
The members of the family and number of meals taken were as follows:

	Meals.
Man 40 years old...	45
Woman 29 years old (45 meals × 0.8 meal of man) equivalent to....	36
Boy 16 years old (45 meals × 0.8 meal of man) equivalent to.......	36
Boy 14 years old (45 meals × 0.8 meal of man) equivalent to.......	36
Boy 9 years old (45 meals × 0.5 meal of man) equivalent to.........	23
Girl 7 years old (45 meals × 0.5 meal of man) equivalent to........	22
Girl 7 years old (45 meals × 0.5 meal of man) equivalent to........	22
Boy 4 years old (45 meals × 0.4 meal of man) equivalent to........	18
Child 16 months old (45 meals × 0.3 meal of man) equivalent to....	14
Total number of meals....................................	252

Equivalent to one man for eighty-four days.

Remarks.—This family lived on a 40-acre cotton farm on a plantation
some 7 miles from Tuskegee in the cotton valley district. Their
one-room cabin was built entirely by themselves and was apparently
warm and comfortable. This last year in addition to their cotton they
raised a small patch of sugar cane and cowpeas. Though in the habit of
mortgaging his crops in past years, the husband is now attempting to
better his condition. The live stock consisted of an old mule, six pigs,
and a hen.

TABLE 61.—*Food materials in dietary No. 141.*

Kind of food material.	Composition.				Weight used.			
	Pro-tein.	Fat.	Carbo-hydrates.	Total cost.	Total food mate-rial.	Nutrients.		
						Pro-tein.	Fat.	Carbo-hydrates.
ANIMAL FOOD.								
Pork:	*Per ct.*	*Per ct.*	*Per cent.*		*Grams.*	*Grams.*	*Grams.*	*Grams.*
Fresh pork [1]	10.6	41.9	....	$1.08	2,720	288	1,140	....
Bacon [1]	7.5	66.1	....	.60	6,095	457	4,029	....
Lard		100.0	....	.39	2,345		2,345	....
Milk	3.3	4.0	5.0	.10	1,135	38	45	57
Total animal food				2.17	12,295	783	7,559	57
VEGETABLE FOOD.								
Corn meal [1]	7.5	4.2	65.9	.18	14,970	1,123	629	9,864
Flour [1]	9.6	.9	78.3	.41	2,720	261	24	2,130
Rice	7.8	.4	79.0	.15	1,360	106	6	1,074
Molasses [1]	4.0	.1	71.8	.29	2,950	30	3	2,118
Sugar			100.0	.06	455			455
Cowpeas [1]	21.6	1.4	61.3	.05	1,985	429	28	1,216
Sweet potatoes	1.5	.6	23.4	.08	4,535	68	27	1,048
Total vegetable food				1.22	28,975	2,017	717	17,905
Total food				3.39	41,270	2,800	8,276	17,962

[1] Average of analyses of similar Alabama foods.

TABLE 62.—*Weights and percentages of food materials and nutritive ingredients used in dietary No. 141.*

Kind of food material.	Food material.	Nutrients.			Food material.	Nutrients.			Cost.
		Protein.	Fat.	Carbohydrates.		Protein.	Fat.	Carbohydrates.	
FOR FAMILY, 15 DAYS.	*Grams.*	*Grams.*	*Grams.*	*Grams.*	*Lbs.*	*Lbs.*	*Lbs.*	*Lbs.*	
Pork, lard, etc.	11,160	745	7,514	....	24.60	1.60	16.60	....	$2.07
Milk	1,135	38	45	57	2.50	.10	.10	0.10	.10
Total animal food ...	12,295	783	7,559	57	27.10	1.70	16.70	.10	2.17
Cereals, sugars, starches..	22,455	1,520	662	15,641	49.50	3.40	1.40	34.50	1.09
Vegetables	6,520	497	55	2,264	14.40	1.10	.10	5.00	.13
Total vegetable food.	28,975	2,017	717	17,905	63.90	4.50	1.50	39.50	1.22
Total food	41,270	2,800	8,276	17,962	91.00	6.20	18.20	39.60	3.39
PER MAN PER DAY.									
Pork, lard, etc	133	9	89	....	.29	.02	.20	....	
Milk	13	....	1	1	.03	....	....	....	
Total animal food ...	146	9	90	1	.32	.02	.20	....	.02¼
Cereals, sugars, starches..	267	18	8	186	.59	.04	.02	.41	....
Vegetables	78	6	1	27	.17	.01	....	.06	....
Total vegetable food.	345	24	9	213	.76	.05	.02	.47	.01¼
Total food	491	33	99	214	1.08	.07	.22	.47	.04
PERCENTAGES OF TOTAL FOOD.	*Per ct.*	*Per ct.*	*Per ct.*	*Per ct.*					*Per ct.*
Pork, lard, etc	27.0	26.6	90.8	....					61.1
Milk	2.8	1.4	.5	.3					2.9
Total animal food....	29.8	28.0	91.3	.3					64.0
Cereals, sugars, starches.	54.4	54.3	8.0	87.1					32.2
Vegetables	15.8	17.7	.7	12.6					3.8
Total vegetable food.	70.2	72.0	8.7	99.7					36.0
Total food	100.0	100.0	100.0	100.0					100.0

TABLE 63.—*Nutrients and potential energy in food purchased in dietary No. 141.*

Kind of food material.	Cost.	Nutrients.			Fuel value.
		Protein.	Fat.	Carbohydrates.	
FOR FAMILY, 15 DAYS.		*Grams.*	*Grams.*	*Grams.*	*Calories.*
Food purchased:					
Animal	$2.17	783	7,559	57	74,740
Vegetable	1.22	2,017	717	17,905	88,350
Total	3.39	2,800	8,276	17,962	162,090
PER MAN PER DAY.					
Food purchased:					
Animal	.02¼	9	90	1	880
Vegetable	.01¼	24	9	213	1,055
Total	.04	33	99	214	1,935
PERCENTAGES OF TOTAL FOOD PURCHASED.					
Food purchased:		*Per cent.*	*Per cent.*	*Per cent.*	*Per cent.*
Animal	64.0	28.0	91.3	.3	45.5
Vegetable	36.0	72.0	8.7	99.7	54.5
Total	100.0	100.0	100.0	100.0	100.0

In the following table a brief summary is given of the food consumed per man per day in the twenty negro dietaries.

TABLE 64.—*Summary of results of negro dietary studies.*

[Food per man per day.]

	Protein.		Fat.		Carbohydrates.		Fuel value.	Nutritive ratio.
Dietary No. 98, farmer:	*Grams.*	*Lbs.*	*Grams.*	*Lbs.*	*Grams.*	*Lbs.*	*Calories.*	
Animal	49	0.11	137	0.30	5	0.11	1.680	
Vegetable	48	.11	11	.02	505	1.11	2,380	
Total	97	.22	148	.32	558	1.22	4,060	1: 9.2
Dietary No. 99, farmer:								
Animal	52	.11	119	.26	65	.11	1,585	
Vegetable	40	.09	5	.01	360	.80	1.685	
Total	92	.20	124	.27	425	.94	3,270	1: 7.7
Dietary No. 100, farmer, summer:								
Animal	2	.01	41	.09			395	
Vegetable	42	.09	16	.04	372	.82	1.845	
Total	44	.10	57	.13	372	.82	2,240	1:11.4
Dietary No. 130, farmer, winter:								
Animal	3	.01	43	.09			410	
Vegetable	32	.07	17	.04	389	.86	1,885	
Total	35	.08	60	.13	389	.86	2,295	1:15.0
Average of dietaries, Nos. 100 and 130:								
Animal	2	.01	42	.09			400	
Vegetable	37	.08	16	.04	380	.81	1.865	
Total	39	.09	58	.13	380	.84	2,205	1:13.2
Dietary No. 101, farmer:								
Animal	27	.06	113	.24	27	.06	1,270	
Vegetable	44	.10	13	.03	405	.89	1,960	
Total	71	.16	126	.27	432	.95	3.230	1: 10.1
Dietary No. 102, sawmill laborer, summer:								
Animal	26	.06	74	.16	26	.06	900	
Vegetable	33	.07	11	.02	403	.89	1,890	
Total	59	.13	85	.18	429	.95	2,790	1: 10.5
Dietary No. 131, sawmill laborer, winter:								
Animal	8	.02	110	.24			1,055	
Vegetable	50	.11	8	.02	390	.86	1,880	
Total	58	.13	118	.26	390	.86	2,935	1:11.3
Average of dietaries Nos. 102 and 131:								
Animal	17	.04	92	.20	13	.03	975	
Vegetable	41	.09	9	.02	396	.87	1,885	
Total	58	.13	101	.22	409	.90	2,860	1:11.1
Dietary No. 103, cotton-plantation laborer:								
Animal	33	.07	261	.57			2.560	
Vegetable	60	.13	22	.05	649	1.43	3,110	
Total	93	.20	283	.62	649	1.43	5,670	1:13.9
Dietary No. 104, cotton-plantation laborer:								
Animal	29	.06	231	.51			2.270	
Vegetable	70	.15	21	.05	666	1.47	3,210	
Total	99	.21	252	.56	666	1.47	5,480	1:12.5

TABLE 64.—Summary of results of negro dietary studies—Continued.

	Protein.		Fat.		Carbohydrates.		Fuel value.	Nutritive ratio.
	Grams.	*Lbs.*	*Grams.*	*Lbs.*	*Grams.*	*Lbs.*	*Calories.*	
Dietary No. 105, farm manager:								
Animal	20	0.05	134	0.29	6	0.10	1,306	
Vegetable	29	.06	4	.01	249	.55	1,175	} 1:11.7
Total	49	.11	138	.30	255	.56	2,535	
Dietary No. 132, farmer:								
Animal	8	.02	74	.16			720	
Vegetable	18	.04	9	.02	225	.50	1,080	} 1:15.9
Total	26	.06	83	.18	225	.50	1,800	
Dietary No. 133, farmer, woman:								
Animal	10	.02	105	.27			1,020	
Vegetable	67	.15	26	.06	649	1.43	3,175	} 1:12.3
Total	77	.17	131	.29	649	1.43	4,195	
Dietary No. 134, farmer:								
Animal	19	.04	114	.25			1,140	
Vegetable	67	.15	27	.06	627	1.38	3,005	} 1:11.0
Total	86	.19	141	.31	627	1.38	4,245	
Dietary No. 135, farmer:								
Animal	7	.02	63	.14			615	
Vegetable	47	.10	22	.19	458	1.01	2,275	} 1:12.1
Total	54	.12	85	.33	458	1.01	2,890	
Dietary No. 136, farmer:								
Animal	12	.03	104	.23			1,015	
Vegetable	37	.08	15	.03	362	.80	1,775	} 1:12.9
Total	49	.11	119	.26	362	.80	2,790	
Dietary No. 137, farmer:								
Animal	1		12	.03			115	
Vegetable	30	.07	15	.03	304	.67	1,510	} 1:11.8
Total	31	.07	27	.06	304	.67	1,625	
Dietary No. 138, farmer:								
Animal	16	.03	107	.24	9	.02	1,100	
Vegetable	36	.08	13	.03	458	1.01	2,145	} 1:14.2
Total	52	.11	120	.27	467	1.03	3,245	
Dietary No. 139, farmer:								
Animal	25	.06	247	.54	1		2,405	
Vegetable	55	.12	22	.05	517	1.14	2,550	} 1:14.1
Total	80	.18	269	.59	518	1.14	4,955	
Dietary No. 140, plantation hand:								
Animal	20	.04	164	.36	1		1,610	
Vegetable	32	.07	18	.04	323	.71	1,625	} 1:14.2
Total	52	.11	182	.40	324	.71	3,235	
Dietary No. 141, farmer:								
Animal	9	.02	90	.20	1		880	
Vegetable	24	.05	9	.02	213	.47	1,055	} 1:13.3
Total	33	.07	99	.22	214	.47	1,935	
Average of above dietaries:								
Animal	19	.04	117	.26	9	.02	1,205	
Vegetable	43	.10	15	.03	427	.94	2,065	} 1:11.8
Total	62	.14	132	.29	436	.96	3,270	
Dietary with minimum protein:								
Animal	8	.02	74	.16			720	
Vegetable	18	.04	9	.02	225	.50	1,080	} 1:15.9
Total	26	.06	83	.18	225	.50	1,800	

TABLE 61.—*Summary of results of negro dietary studies—Continued.*

	Protein.		Fat.		Carbohydrates.		Fuel value.	Nutritive ratio.
Dietary with maximum protein:	Grams.	Lbs.	Grams.	Lbs.	Grams.	Lbs.	Calories.	
Animal	29	0.06	231	0.51			2,270	
Vegetable	70	.15	21	.05	666	1.47	3,210	} 1:12.5
Total	99	.21	252	.56	666	1.47	5,480	
Dietary with minimum energy:								
Animal	1		12	.03			115	
Vegetable	30	.07	15	.03	304	.67	1,510	} 1:11.8
Total	31	.07	27	.06	304	.67	1,625	
Dietary with maximum energy:								
Animal	33	.07	261	.57			2,560	
Vegetable	60	.13	22	.05	649	1.43	3,110	} 1:13.9
Total	93	.20	283	.62	649	1.43	5,670	

DISCUSSION OF RESULTS.

It is evident that the families in and near the village of Tuskegee whose condition and food consumption are here reported do not fairly represent the average plantation negroes who make up the larger part of the colored population of the black belt. They may represent even less accurately the colored population of the town. They have been influenced more or less by the Tuskegee Institute. It seems probable that the families on the plantation at a distance from Tuskegee whose dietaries are reported here more nearly represent the average plantation negroes. The statistics are in both cases limited in number, and it is evident that it would be unwise to make broad and definite generalizations regarding the food of the negro in the Southern States, its effect upon his physical, mental, and moral character and efficiency, and the means that should be adopted for its improvement. These investigations were intended rather as preliminary work in a field where an extended and accurate survey is needed.

Such information as could be obtained from conversation with people who are familiar with the negro population of the Southern States and personal observations leave the impression that the condition of the average negro family resembles Nos. 100–101, rather than the more thrifty families like Nos. 98 and 99. These, and especially No. 98, are evidently exceptions and presumably quite rare ones. The statements of Mr. Washington, of Tuskegee, as well as those of other gentlemen who have had large opportunities for observation, indicate that the one-room cabin is the common habitation and that the ordinary furnishings and ordinary diet are decidedly inferior to the average of those here reported.

It is evident that while the diet of the negro in the South is a very important factor of his character and condition, its effect is hardly to be separated from that of the other conditions of his existence. Diverse statistical and sociological data will be necessary before all the desired conclusions can be reached.

65

Much may be learned, however, from such inquiries as these. Further inquiries are now in contemplation, and it is believed the data here recorded can be best discussed when more material is available. Meanwhile, a brief comparison of the negro dietaries with dietaries of people of various conditions and with so-called dietary standards may not be without interest. Such a comparison is made in Table 65.

TABLE 65.—*Comparison of Tuskegee negro dietaries with other dietaries in this country and in Europe and with dietary standards.*

[Quantities per man per day.]

Dietaries.	Protein.		Fat.		Carbohydrates.		Fuel value.	Nutritive ratio.
	Grams.	Lbs.	Grams.	Lbs.	Grams.	Lbs.	Calories.	
NEGROES NEAR TUSKEGEE.								
No. 137, farmer	31	0.07	27	0.06	304	0.67	1,625	1:11.8
No. 132, farmer	26	.06	83	.18	225	.50	1,800	1:15.9
No. 141, farmer	33	.07	99	.22	214	.47	1,935	1:13 3
No. 100, farmer, summer	44	.10	57	.13	372	.82	2,240	1:11.4
No. 130, farmer, winter	35	.08	60	.13	389	.86	2,295	1:15.0
Average	39	.09	58	.13	380	.84	2,265	1:13.2
No. 105, farm manager	49	.11	138	.30	255	.56	2,595	1:11.7
No. 136, farmer	49	.11	119	.26	362	.80	2,790	1:12.9
No. 102, sawmill laborer, summer	59	.13	85	.18	429	.95	2,790	1:10.5
No. 131, sawmill laborer, winter	58	.13	118	.26	390	.86	2,932	1:11.3
Average	58	.13	101	.22	409	.90	2,860	1:11.1
No. 145, farmer	54	.12	85	.33	458	1.01	2,890	1:12.1
No. 101, farmer	71	.16	126	.27	432	.95	3,230	1:10.1
No. 140, plantation hand	52	.11	182	.40	324	.71	3,235	1:14.2
No. 138, farmer	52	.11	120	.27	467	1.03	3,245	1:14.2
No. 99, farmer	92	.20	124	.27	425	.94	3,270	1: 7.7
No. 98, farmer	97	.22	148	.32	558	1.22	4,060	1: 9.2
No. 133, farmer—woman	77	.17	131	.29	649	1.43	4,195	1:12.3
No. 134, farmer	86	.19	141	.31	627	1.38	4,235	1:11.0
No. 139, farmer	80	.18	269	.59	518	1.14	4,955	1:14.1
No. 104, cotton plantation laborer	99	.21	252	.56	666	1.47	5,480	1:12.5
No. 103, cotton plantation laborer	93	.20	283	.62	649	1.43	5,670	1:13.9
Average of all	62	.14	132	.29	436	.96	3,270	1:11.8
POOR PEOPLE, UNITED STATES.								
25 families in poorest part of Philadelphia:								
Smallest dietary, negro	66	.15	68	.15	181	.40	1,630	1: 5.4
Largest dietary, German	202	.45	266	.45	608	1.34	5,235	1: 5.3
Average	109	.24	108	.24	435	.96	3,235	1: 6.2
26 families in poorest part of Chicago:								
Smallest dietary	86	.19	100	.22	213	.47	2,195	1: 4.6
Largest dietary	168	.37	204	.45	626	1.38	4,950	1:11.3
Average	119	.26	141	.31	398	.88	3,425	1: 6.0
PEOPLE IN MORE COMFORTABLE CIRCUMSTANCES, UNITED STATES.								
Farmer, Connecticut	79	.17	117	.26	354	.78	2,865	1: 7.8
Farmer, Connecticut	104	.23	156	.34	494	1.09	3,900	1: 8.1
Average 5 dietaries, farmers in Connecticut	92	.20	114	.25	483	1.06	3,420	1: 8.1
Carpenters, Connecticut	105	.23	136	.30	362	.80	3,185	1: 7.4
Timber, Indiana	90	.20	134	.30	408	.90	3,285	1: 7.9
Boarding house, well-paid machinists, etc., Connecticut	103	.23	152	.34	401	.88	3,490	1: 7.3
Mechanic, Tennessee	110	.24	210	.46	412	.91	4,090	1: 8.1
Average 9 dietaries of mechanics, etc	105	.23	152	.34	420	.92	3,570	1: 7.3

12246—No. 38——5

66

TABLE 65.--*Comparison of Tuskegee negro dietaries with other dietaries in this country and in Europe and with dietary standards*—Continued.

Dietaries.	Protein. Grams.	Lbs.	Fat. Grams.	Lbs.	Carbohydrates. Grams.	Lbs.	Fuel value. Calories.	Nutritive ratio.
PEOPLE IN MORE COMFORTABLE CIRCUMSTANCES, UNITED STATES—continued.								
Boarding house, Lowell, Mass., boarders operatives in cotton mills	132	0.29	200	0.44	594	1.21	4,650	1: 7.6
Average 29 dietaries of people at active exercise, mechanics, etc., in Massachusetts and Connecticut	154	.34	227	.50	626	1.38	5,275	1: 7.5
PROFESSIONAL MEN.								
Average of 9 dietaries	104	.23	122	.27	428	.94	3,315	1: 6.8
COLLEGE STUDENTS' BOARDING CLUBS, UNITED STATES.								
Average of 15 dietaries	108	.24	148	.33	460	1.01	3,700	1: 7.4
POOR PEOPLE SCANTILY NOURISHED, EUROPEAN.								
Working people, Saxony, average 13 dietaries	69	.15	45	.10	384	.85	2,275	1: 7.0
Mechanics, laborers, beggars, etc., Naples, Italy, average 5 dietaries	76	.17	38	.08	396	.87	2,290	1: 6.3
Farm laborer, Saxony, food mainly vegetable	89	.18	37	.08	504	1.11	2,740	1: 7.4
Farm laborer, Prussia, food mainly vegetable	83	.18	17	.04	373	1.26	2,845	1: 7.4
PEOPLE IN MORE COMFORTABLE CIRCUMSTANCES, AT MODERATE WORK, EUROPEAN.								
Bavaria, average 11 dietaries of carpenters, coopers, and locksmiths	122	.27	34	.08	570	1.26	3,150	1: 5.3
Peasants near Moscow	129	.28	33	.08	589	1.30	3,250	1: 5.1
Average 5 dietaries of farm laborers, Bavaria	137	.30	55	.12	542	1.19	3,295	1: 4.9
Average 6 dietaries of mechanics, etc., southern Sweden	134	.30	79	.17	523	1.15	3,435	1: 5.2
Peasant farm laborer, Italy	118	.26	65	.14	628	1.38	3,665	1: 6.6
PEOPLE AT ACTIVE EXERCISE, EUROPEAN.								
Average 5 dietaries of machinists, etc., southern Sweden	189	.42	110	.24	714	1.57	4,725	1: 5.1
Farm laborers, Austria, diet, corn meal and beans	159	.35	62	.14	977	2.15	5,235	1: 7.0
Javanese in Java village, World's Fair, Chicago	66	.15	19	.04	254	.56	1,490	1: 4.5
United States Army rations	120	.26	161	.36	454	1.00	3,850	1: 6.8
DIETARY STANDARDS.								
European:								
Woman at moderate work	92	.20	44	.10	400	.88	2,425	1: 5.5
Man at moderate work	118	.26	56	.12	500	1.10	3,055	1: 5.3
Man at hard work	145	.32	100	.22	450	.99	3,370	1: 4.7
American:								
Woman with light muscular exercise	90	.20					2,400	1: 5.5
Woman with moderate muscular work	100	.22					2,700	1: 5.6
Man without muscular work	100	.22					2,700	1: 5.6
Man with light muscular work	112	.25					3,000	1: 5.5
Man with moderate muscular work	125	.28					3,500	1: 5.8
Man with hard muscular work	150	.33					4,500	1: 6.3

In the above table the figures for mechanics' families in New Jersey, Indiana, and Tennessee are from studies made in connection with the general inquiry of which the present one forms a part. The rest of the figures in the table are taken from a summary of results of investigations of dietaries in different parts of the world.[1] They are selected as illustrating as well as possible with so limited a number of cases the differences in the kinds and amounts of nutriment in the food of people of different places and classes. Unfortunately the amount of available information upon this larger subject is quite limited. The summary just referred to includes the studies of less than 400 different dietaries. They will, however, suffice for general comparison.

The dietary standards in the article are intended to represent the quantities of nutrients appropriate to an average person in each of the classes referred to. They are based upon two classes of data, namely, the amounts of nutrients actually contained in the food of well nourished people and the results of more or less accurate experiments upon the nutrients of single individuals also well nourished. The European authorities quoted are those whose judgment is commonly accepted by their fellow specialists throughout the world. The American estimates have been made more liberal on account of the results of late investigations in this country, which were not available to the European investigators and which imply a larger food consumption by people of the laboring classes in this country than by those of corresponding classes in England and especially on the Continent of Europe.[2]

Bearing in mind that these so-called dietary standards are not in any way absolute, but simply represent the best knowledge of the subject we have to-day, the figures for the nutrients in the food of a man at moderate muscular work may be taken as a basis for comparison. The two principal items are the quantities of (1) protein, the so-called "flesh formers," and (2) fuel values. According to the German standard of Voit, a man engaged in moderately hard manual labor—a carpenter, mason, or day laborer, for instance—ought to have 0.26 pound of protein to form blood, muscle, bone, and other nitrogenous parts of the body, and thus make up for the constant wear and tear of the bodily machine; and, in addition, enough of the fuel ingredients, fats and carbohydrates, to furnish 3,055 calories of energy to be transformed into heat, muscular power, and other forms of energy needed to keep the bodily machine in successful operation. The nutritive ratio of such a dietary, i. e., the proportion of protein to fuel ingredients (reckoning 1 part by weight of fats as equivalent in fuel value to $2\frac{1}{4}$ parts of carbohydrates) would be 1 : 5.3.

The American standard, suggested above, for a man of the same class assumes that he does, on the whole, rather more work and needs

[1] U. S. Dept. Agr., Office of Experiment Stations Bul. 21.
[2] For discussion of this subject see U. S. Dept. Agr., Office of Experiment Stations Bul. 21, p. 206.

nutrients in his daily food sufficient to furnish 0.28 pound of protein and 3,500 units of energy with a nutritive ratio of 1 : 5.8.

The estimates for men at more active muscular work are considerably larger, and for those engaged in occupations which require but little muscular exercise, as is apt to be the case with professional and business men, the needed amounts of protein and energy are assumed to be smaller.

It will be observed that the figures of Table 65 for the food consumption of well-fed and well-to-do people generally in the United States and those of similar classes in Europe agree more or less closely with the dietary standards. The negro dietaries show on the average a liberal allowance of fuel ingredients in the food as measured by the fuel values. But the quantities of protein in the negro dietaries are extremely small, in general from one-half to two thirds the amounts which the standards call for and which are actually found in the food of well-to-do and well-nourished people of different classes in the United States and in Europe. The nutritive ratios of the negro dietaries are very wide as compared with those of both the dietary standards and the actual dietaries of people who are ordinarily assumed to be well nourished. Thus in the food of well-to-do and well-paid mechanics in Germany the quantities of protein average about 0.27 pound, being generally larger in the food of those who are engaged in more active manual labor. The quantities in the American dietaries with which those of the negroes are compared are on the whole rather larger than the foreign, but in the negro dietaries the range is only from 0.10 to 0.22 pound. The fuel values in the European dietaries range from 1,650 to 5,235, in the American dietaries from 1,630 to 5,285, and in the Tuskegee dietaries from 1,625 to 5,670.

Comparing these negro dietaries with other dietaries and dietary standards it will be seen that—

(1) The quantities of protein are very small; roughly speaking, the food of these negroes furnished one-third to three-fourths as much protein as are called for in the current physiological standards and as are actually found in the dietaries of well-fed whites in the United States and well-fed people in Europe. They were, indeed, no larger than have been found in the dietaries of the very poor factory operatives and laborers in Germany and the laborers and beggars in Italy.

(2) In fuel value the negro dietaries compare quite favorably with those of well-to do people of the laboring classes in Europe and the United States.

(3) The marked peculiarity of the negro dietaries, namely, their lack of protein, is shown in the nutritive ratios. While the proportion of protein to fuel ingredients in the dietary standards and in the food of well-fed wage workers ranges from 1 : 5 to 1 : 7 or 8, and is about 1 : 5.5 or 1 : 6 in the dietary standards, the nutritive ratio of the negro dietaries range from 1 : 7 to 1 : 16. Leaving out two quite exceptional cases, the lowest was 1 : 10 and the average 1 : 11.8.

The following diagram [1] shows the relative amounts of food materials and the nutritive value of the actual daily dietaries of a negro field laborer and farmer as compared with the nutritive value which a properly balanced dietary should have:

Actual daily dietaries of negro field laborer and farmer compared with a well-balanced standard dietary.

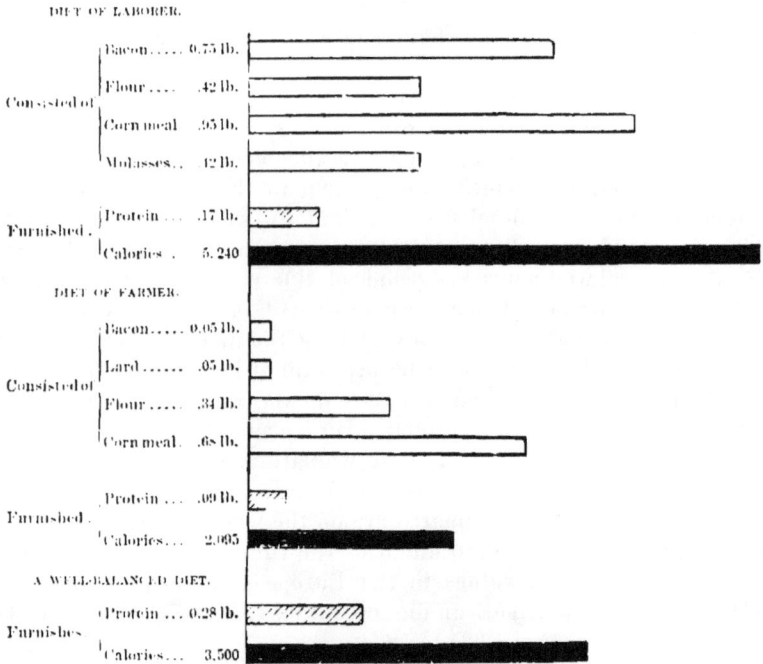

DIET OF LABORER.

Consisted of	Bacon..... 0.75 lb.	
	Flour .... .42 lb.	
	Corn meal .95 lb.	
	Molasses.. .12 lb.	
Furnished .	Protein .... .17 lb.	
	Calories . 5,240	

DIET OF FARMER.

Consisted of	Bacon..... 0.05 lb.	
	Lard...... .05 lb.	
	Flour ..... .34 lb.	
	Corn meal .68 lb.	
Furnished .	Protein ... .09 lb.	
	Calories... 2,095	

A WELL-BALANCED DIET.

Furnishes	Protein ... 0.28 lb.	
	Calories... 3,500	

The field laborer was both underfed and overfed, since the food contained too little protein and too much fuel value. The farmer was underfed; the food had only one-third of the protein and two-thirds the fuel value needed.

This diagram is adapted from a wall chart used at Tuskegee Institute and exhibited in the negro building at the Atlanta Exposition in 1895.

www.ingramcontent.com/pod-product-compliance
Lightning Source LLC
Chambersburg PA
CBHW020239090426
42735CB00010B/1765